TEAMING A PRODUCT AND A GLOBAL MARKET

A
Canadian
Marconi
Company
Success
Story

Graham Gibbs

American Institute of Aeronautics and Astronautics, Inc.
1801 Alexander Bell Drive
Reston, Virginia 20191

Publishers since 1930

American Institute of Aeronautics and Astronautics, Inc., Reston, Virginia

Library of Congress Cataloging-in-Publication Data

Gibbs, Graham, 1945–
   Teaming a product and a global market : a Canadian Marconi Company success story / Graham Gibbs.
      p.   cm.
   Includes index.
   1. Canadian MArconi Company—History.   2. Aircraft supplies industry—Canada—History.   3. Aeronautical instruments industry—Canada—History.   4. Avionics—History.   5. Airplanes—Electronic equipment.   6. Airplanes—Electronic equipment.   I. Title.
HD9711.2.C24C364   1997   338.7'629135'0971—dc21   97-18855
ISBN 1-56347-225-2

Cover design by Christine Moe

Copyright © 1997 by the American Institute of Aeronautics and Astronautics, Inc. All rights reserved. Printed in the United States of America. No part of this publication may be reproduced, distributed, or transmitted, in any form or by any means, or stored in a database or retrieval system, without the prior written permission of the publisher.

In Memory of Christina

This book is for Zanna and Amy.
They continue to enrich and make my life complete.

This book is also for all the people on the Marconi team,
who searched for and found excellence,
and for Al Browne;
with his premature death we lost a friend and a product
champion. If Al had lived, he would have written about another
Canadian Marconi success story.

CONTENTS

Preface vii
Acknowledgments ix

Part One: The Beginning of an Era 1
1. Product Champions 3
2. Pioneering a Product and a Market 17
3. I Join the Canadian Marconi Company 29
4. The Search (for a Customer) Continues 39
5. Where Do We Go from Here? 51
6. A Customer at Last 59
7. A Second-Generation Omega Navigation System 71
8. Looking Back 77

Part Two: A Global Market 85
9. "I'm From the Government and I'm Here to . . ." 87
10. Pan American World Airways, Thank You! 99
11. "One Sure Customer Can Lead to Another" 113
12. Building Our Market 131
13. An Organization to Fit the Market 143
14. Commitment to a Market 157
15. Taking the Initiative to Expand the Market 165
16. "Profit" Is Not a Dirty Word 181
17. Maintaining Momentum 193
18. Putting It All Together 213
19. They Did Not Become Complacent 231

Epilogue: Was It Worth It? 241

Index 245

Preface

I wrote this book for several reasons, some were personal. Foremost, the story is worth telling. Succeeding in a high-technology export market in competition with some of the largest companies makes our story unusual. Achieving a global leadership position without a domestic market is remarkable.

Teaming a Product and a Global Market is about transitioning into a commercial market, exporting, and launching new products. This is also a story about how a small team at the Canadian Marconi Company achieved world leadership in a global market.

I describe the lessons we learned on the road to success, why we succeeded, and why, sometimes, we failed. Though this story spans the two decades of the 1970s and 1980s our experiences remain relevant. We employed concurrent engineering, practiced empowerment, organized ourselves as integrated product teams, and much more that is in vogue today. The difference for us was that our business practices seemed so logical that we did not assign names to them.

I hope this book will be instructive, as well as entertaining as leisure reading. Also, I hope it will encourage business people to compete in the global market.

Central to this story is an aircraft navigation system that uses signals from eight government-owned stations around the world. The aviation community knows it as the Omega System.

The Canadian Marconi Company developed its first aircraft Omega Navigation System while the government of the United States of America was setting up the ground stations. Part of this story is our recognition that if we wanted to sell our own product, we had to help promote this new technique in navigation.

Because the Omega System was a new concept in radio navigation, we had to overcome many technical and nontechnical problems. The causes of most of those problems were beyond our control; however, our success relied on our ability to find solutions.

My decision to leave the Canadian Marconi Company was the hardest decision I have had to make. After 27 years with Marconi companies in England and Canada, I felt the need for a new challenge and an adventure. For a while I had held a desire to represent Canada in some capacity. The space business has always fascinated me. Then the Canadian Space Agency offered me a job that satisfied my three ambitions. Still, it was hard to leave Canadian Marconi.

I have great respect and admiration for the company and its people. I consider myself to be a team player. Because teamwork cultivates loyalty, the hardest part of my decision was the feeling that I was being disloyal.

My respect for the Canadian Marconi Company has not diminished. It is a company of innovators with the courage to be creative—in the laboratory and in the market. This is our story.

Graham Gibbs
June 1996

Acknowledgments

Throughout this book I have recalled those people who had the most influence on me when I worked with the team at the Canadian Marconi Company. I hope, should they read this book, they will agree I have accurately portrayed their roles.

There were many others on the team. I have not deliberately omitted anyone. I owe a deep debt of gratitude to all the people of Canadian Marconi and to many others in the international aviation community. This period in our life was challenging and fun.

I would be extremely remiss if I did not emphasize the role of Christina. She was a constant source of nurture. She supported me while I was busy developing a career. She helped me immeasurably during the hard times but shared the joy of the good times. Above all else, she was the mainstay of our family, the parent who was always there. Our daughters, Zanna and Amy, are a constant joy to me and have given of themselves as much and perhaps more than they have received.

Only people who have had a book published can understand the true feeling behind the sentence in an Acknowledgments that contains these words: "... and I would like to thank my wife because without her understanding this book would never have been completed." When I started this book, I had no idea that I would spend nearly 1500 hours just preparing the manuscript for a professional editor. It is with deep affection, and knowledge gained the hard way, that I remember Christina for her support. She did not complain during the two years I spent most evenings and many weekends alone while I pursued my ambition to write this book. Zanna and Amy also were tolerant. Many times they completed their homework on a computer that should have been in a museum while I commandeered the newer addition to the family's workroom.

I want to also single out several other people who helped me during our search for excellence.

Kieth Glegg was vice president of the Canadian Marconi Company, Avionics Division, during my first three years with the company. He had vision and an understanding of the marketplace. Also, he was not above helping in the trenches, when necessary. As a young aspiring marketeer in a highly competitive business, I learned much from Kieth.

Ervin Spinner has been a manager, mentor, and, I hope he will agree, friend, as well as another veteran Marconi employee. Throughout my career, and while writing this book, Ervin provided me with invaluable guidance. He provided the vision for our success. Many went to him for solutions, that is not his style. Ervin provided us with an intellectual analysis of our problems that was sufficient for us to develop our own solutions. This effective management technique helps people to think for themselves, while also providing counsel for today's problem.

Peter and Lise Gasser and Tony and Liliane Sayegh are business colleagues and close friends. They have contributed to my life in many ways. At Canadian Marconi, we worked as a team: Peter was the engineer and manager, Tony was our contracts expert, and I was the marketeer.

When John Simons became general manager of the Avionics Division in 1976, he brought fiscal control and calculated planning to our business. After several years of courageous new product development, we needed to be more disciplined. John, the consummate professional, helped us all to chart and to steer a clear course into the future, a future that brought many rewards and sustained profitability.

John and Maureen Rogers are the two people, outside my family, who have had the most influence on me. They would undoubtedly deny this accolade. I first met John and Maureen at a scuba diving club in Montreal. I knew John as a beer peddler because he often described himself that way, even though he was a president and later deputy chairman of the Molson Group of Companies. What I admire are his leadership qualities, his thoughtfulness for all who come into contact with him, his dedication, his willingness to volunteer his time for many causes, and his ability to inspire. John taught me the first rule of business: no matter what your position in a company you are first its salesperson. He and Maureen have remained important close friends.

Patricia Carnes, my administrative assistant in Washington, DC for five years, was a superb administrator and a valued friend. Though she and her husband have now moved to Florida, she continues to be a source of help and encouragement.

Often it is a small incident or chance remark that sets off a chain of events. Before writing this book, I wrote two short novels. One was about my experiences at a British boarding school. Writing that book brought back many memories and the incentive to contact after 30 years people from my youth. During a visit with my English teacher, Daisy Hardy, we talked about my world travels for Canadian Marconi. "Oh, Graham!" she finally exclaimed, "how exciting. You should write a book." It was the only encouragement I needed. I hope, should she read this book, she will accept it as atonement for all the bad essays I handed to her so many years ago.

Part One

The Beginning of an Era

1

Product Champions

Opportunity is missed by most people because it is dressed in overalls and looks like work.
—Thomas Edison

I knew that we had a formidable task ahead of us when a voice from the back of the room interrupted my sales presentation with these words: "Macaroni . . . wasn't he the Italian guy who invented the radio?"

At that point I realized that not only did we have to promote the product, we had to promote the company. Perhaps we even had to promote Canada. No, I did not work for an Italian spaghetti factory that also produced sophisticated electronics equipment for aircraft. I worked for the Canadian Marconi Company, a Canadian corporation whose major shareholder is the General Electric Company of the United Kingdom.

Guglielmo Marconi was an Italian by birth. However, it was in England that he perfected his experiments in the transmission of electrical signals without the use of wires connecting the transmitter with the receiver. The radio as we know it today, utilizing Marconi's techniques, came a little later. From those tentative beginnings he built a formidable electronics company known, I thought, to everyone.

Canadian Marconi Company

There are moments in history when challenges occur of such a compelling nature that to miss them is to miss the whole meaning of an epoch.
James Michener

Heard on a headset, Dec. 12, 1901:
Click___Click___Click

One year later:

> HIS MAJESTY THE KING. MAY I BE PERMITTED BY MEANS OF FIRST WIRELESS MESSAGE TO CONGRATULATE YOUR MAJESTY ON SUCCESS OF MARCONI'S GREAT INVENTION CONNECTING CANADA AND ENGLAND. MINTO.
>
> Lord Minto, Governor-General of Canada

On July 20, 1897, Guglielmo Marconi established the Wireless Telegraph and Signal Company Limited to produce and to operate the short-range ship-to-shore radio equipment that he had invented.

On Dec. 12, 1901, at 12:30 p.m. Newfoundland time, Guglielmo Marconi heard three short clicks among the static in a headset. To confirm his remarkable achievement Marconi passed the headset to his assistant George Kemp. The three clicks, the Morse code letter S, were repeated several times. The world's first transatlantic wireless communication had taken place. The transmitter was located at Poldhu in the west of England. The receiver was on a promontory overlooking Saint John's, Newfoundland, Canada.

Earlier in the same year, on Sept. 17, the worst gale in living memory struck Cape Cod in Massachusetts and blew down Marconi's first antenna in North America. If the gale had not struck, Canada may not have been selected for the site of Marconi's first permanent North American station.

The first transatlantic communication without wires, wireless communication, in 1901 heralded a new era in communications. This was the birth of radio. Its parent was Guglielmo Marconi, and his achievement altered the course of communications history, the growth of which has no parallel.

The success of the transatlantic transmission led to the creation, in 1902, of the Marconi Wireless Telegraph Company of America, of which Canadian Marconi Company was a part.

After its formation, the parent company expanded dramatically, and its range of products included just about every conceivable industrial electronics product. The company also produced consumer radios and television sets but dropped those products in the 1950s to concentrate on sophisticated military and commercial electronics products.

In 1946 the company amalgamated with the English Electric Company. Later English Electric–Marconi was amalgamated with the General Electric Company of England. Today the corporation is a large international conglomerate.

Guglielmo Marconi originally set up the Canadian Marconi Company, in 1903, as a communications broadcasting company, but it soon became necessary for the company also to produce the broadcasting equipment, and so, Canadian Marconi became both an operator of radio stations and a manufacturer of radio equipment. That dual role, which later expanded to include Canada's first television station, continued until 1972. Then the radio and television stations were sold, but by this time the Canadian Marconi Company was offering a range of communications and aircraft electronics products.

When I joined the company in 1972, the headquarters and manufacturing plant were located in Montreal, where they remain today. Expansion into other cities and the United States would occur later. The company employed 2500 people in three major operating divisions that specialized in marine and land communications equipment, defense communications equipment, and electronics equipment for aircraft, commonly called avionics. The word avionics is derived from aviation electronics.

Over the years the company has undergone some restructuring as new divisions have been created and one or two combined with others. My story occurred within the Avionics Division, which has remained a dominant operating division since its formation in the 1960s.

In the 1950s and 1960s the Avionics Division of Canadian Marconi had gained a worldwide reputation as a leader in aircraft Doppler Navigation Systems. Those systems rely on the transmission and reception of an electronic signal from equipment installed in the aircraft. Canadian Marconi sold that equipment in large quantities to military agencies throughout the world and in particular to the U.S. Army and Air Force.

Canadian Marconi was in the process of developing an airline market for the Doppler Navigation Systems when it won a large long-term military contract in the United States. The Vietnam War started, which meant more orders. At this point Kieth Glegg, the vice president, and others decided to concentrate exclusively on the military market, even though a few airlines had bought systems. Some of them saw that decision as a lack of commitment by Canadian Marconi. That perception proved to be a major hurdle that we had to overcome some 20 years later when we launched a concerted effort to reestablish an airline customer base. That effort is part of this story.

Until 1968 the Avionics Division had a conventional, functional organization. There was a central manufacturing organization, an engineering organization, a central marketing group, and so on. The details of the pre-1968 organization are not important; the post-1968 organization is important.

In 1968 Kieth Glegg initiated a significant change in the organizational structure of the Avionics Division. Many people wondered if the division could endure such a change in the way the business was managed. Twenty-five years later the organization is substantially the same—it withstood the test of time. The division prospered under that new organization, although I believe that it benefited from some of the fine-tuning that we introduced during the 1970s and 1980s as we searched for and achieved success in a broader business environment.

By 1968 the Avionics Division had several very large contracts that were being executed within a functional organization. Within that structure no one department felt responsible for the overall execution of a contract, only one part of it. The engineering department worried about the engineering aspects and left the manufacturing problems to the manufacturing department. Manufacturing, in turn, did not busy itself with the contractual niceties; that, after all, was the job of the contracts department.

A customer had to talk to a variety of people to get answers to the usual host of questions that follow a contract. Kieth felt that the Avionics Division could better serve its customers if the authority for a contract was held farther down the line. His solution was a matrix organization. So innovative was that concept at the time that the reorganization of the Avionics Division became the subject of a Harvard Business School case study.

Kieth's matrix organization initially centered around a program manager. The program manager was responsible for a single long-term contract for a particular product. To help with the task, an engineering team, contracts administration staff, manufacturing personnel, customer support personnel, and technical draftsmen were assigned to the program manager, who was thus given direct control over all the disciplines required to execute a contract. The program manager, in effect, operated a mini operating division where he or she had profit and loss responsibility for a contract that often amounted to several million dollars. The contracts, with follow-on orders, would often last several years.

The matrix aspect of the organization came about because there was still a division-level manufacturing manager, an engineering manager, a contracts manager, and a product support manager. As a result, the personnel assigned to a program manager had two supervisors: the program manager, who directed day-to-day activities, and the organizational chief, who managed the longer-term objectives of the functional organization.

After two years Kieth decided to reorganize the matrix organization beyond narrowly defined programs or single contracts. As a result, he introduced a product management concept; product managers were given the responsibility for entire product lines or technologies. In that capacity they directed the initial development of a product, managed it through production, found customers for the product line, sold it, and supported the product once it was delivered. In cases where a product line had several large single contracts, program managers were assigned to those contracts, and they reported to the product manager.

Other than the size of the business, the biggest difference between the responsibilities of a product manager and a program manager was that the product manager had a responsibility to market the product line. To accomplish that task the product manager was assigned applications engineers, who were marketing professionals with a strong engineering background. Their job was not only to market the product line but also to work with potential customers to determine the right configuration of system for their particular needs.

One of our biggest assets was our willingness to work with customers and customize our products to meet their needs. That was no mean feat because most avionics are complex and sophisticated products, built to exacting standards. With such products it is difficult to be both flexible and profitable. Ideally, a company wants to build many of exactly the same item and not a multitude of variations. We managed to find a way to be flexible and profitable. At one point we were offering more than 40 variations of the product that is central to this story, with all variations coming off one assembly line. We would have gone bankrupt within a week if we had tried to set up 40 different assembly lines.

Of course, the product managers could not be given total autonomy. A system of checks and balances existed, and as in any organization, certain tasks required a more centralized approach. Long-term planning, staff assignments, capital equipment purchases, manufacturing plans, and bid strategies, including pricing, to name only a few, all required a centralized management approach.

Group managers helped to coordinate these activities and those of other product lines. The group managers had three or four product managers reporting to them. When I was working in the division, the senior management team consisted of the product group managers, the group managers of the functional organizations, and the vice president, about seven individuals altogether.

Shortly before I finished this book and long after I had written this chapter, an event occurred that made me realize that mistakenly I had taken for granted one other aspect of that matrix organization. Consequently, I found it necessary to return to this chapter.

I was participating in a meeting where several major aerospace companies were making presentations. At one point the participants discussed the benefits of Total Quality Management, the name given to a set of management techniques that now have worldwide attention.

One company reported on the cost savings it was projecting by employing concurrent engineering. I was not familiar with the term and asked for an explanation. To my surprise the answer was a strategy for engineering that we had practiced at Canadian Marconi and one that I had taken for granted. It was not a new revelation after all but an approach to engineering that now had a name; apparently it was new to some.

To bring a product to market, four basic types of engineering have to be carried out: design and development—designing the basic product; producibility engineering—influencing the design to ensure that it can be manufactured efficiently; maintainability engineering—incorporating design features to ensure that the product can be efficiently maintained and serviced; and, finally, reliability engineering—incorporating technologies and design features that contribute to the eventual reliability of the product.

At Canadian Marconi, we always strived to coordinate and carry out those various engineering disciplines at the same time the product was being designed. Our technique now goes by the name concurrent engineering. I am not trying to suggest that we were perfect in applying all the engineering disciplines at the same time. However, we did make a very conscious attempt. It just seemed to us to be a sensible approach to develop a product efficiently.

Apparently, others carry out those engineering tasks in a sequential manner. That means they have to keep going back to change the basic design when it does not work as expected in production or with a customer, which is an expensive approach.

When Kieth Glegg introduced his matrix organization, he succeeded in placing the authority where it belonged, with the people who really understood the product and its customers. He had given those people enough authority and the multidisciplined staff they needed to fulfill their mandate.

It is a cause of wonder to me that as Canadians we seldom acknowledge our accomplishments. Canadians have a reputation for being modest. Making sure that the world knows about our achievements is not bragging—it is good marketing. Even our own citizens are poorly informed about Canadian achievements.

Some achievements, such as the U.S. Space Shuttle Remote Manipulator System, the Canadarm, receive a lot of publicity. Sometimes, though, one would think that is all we have accomplished. The Canadarm is an incredible technological achievement, but there are many other equally significant Canadian successes. My purpose, in writing this book, is to recount one success story and the lessons we, at Canadian Marconi, learned as we searched for and achieved excellence in our field of endeavor.

Successful Canadian companies have developed exceptional export marketing skills because the Canadian home market is too small to sustain their businesses. Canadian aerospace companies, and others, have to compete with the world's largest and best. It is not widely acknowledged that those Canadian companies not only hold their own with those giants but also oftentimes beat them. Canadian aerospace companies export more than 70% of what they produce, whereas, in 1989, U.S. aerospace companies exported only 25% of their total sales.

The successful Canadian aerospace companies have learned to select and to develop products for niche markets. Canadian designs are innovative and frequently use the latest or evolving technologies and product concepts.

Canadian Marconi fits that description of success. Customers acknowledge its products as being of world class. The company exports nearly 90% of its products to over 100 countries.

Export sales, success, profitability, and world leadership are earned, not given. Many lessons must be learned on the way to success. Recognizing those lessons, remembering them, accepting and learning from failures and continuously applying the lessons learned are key ingredients to success. The attainment and continuance of success require tenacity, commitment, and a willingness to listen and to adapt. Above all, success requires teamwork.

We Were a Team . . .

Whatever I may tell myself about my own work, I know, better than anyone else possibly can, how little I could have done alone.
John Alvin Pierce

Many people believe that products succeed only when they are championed by someone, a product champion. While one individual may have championed the original idea, it takes teamwork to design a product, to launch it into the market, and to achieve sustained market success. We had product champions.

Throughout this account I have liberally used the noun "we" and have used "I" only for those occasions when I write about my lone participation in an event or when I wish to relate a personal thought. Nonetheless, I write from my own perspective of the events as experienced through my involvement in marketing, not engineering or manufacturing. Others may have a different view of why we succeeded and sometimes failed.

We sometimes hear marketing professionals argue that they deserve salary increases because they secured x million dollars of business during the last year. The inference is that the marketing persons secured that business alone. This is not true. They may have identified the most likely customers, developed a marketing plan, convinced the organization to support the plan, and made many trips to visit potential customers. However, that was their job. The successes that they claim occurred because the company management gave them the support they needed, the engineers developed the product, the manufacturing department built it, the quality assurance department made sure the product was built correctly, and the field engineers supported the product. Also, the accounts department arranged for the marketing professionals to receive salaries regularly. They did not do it all alone. It was a team effort.

Company presidents may expound on how they restored profitability and saved their companies from almost certain bankruptcy. While they undoubtedly provided the vision and leadership to achieve those profits, they also must have had the support of a dedicated team.

Recognizing the team aspects of any business or a group recreational activity is a key ingredient to a company's and one's personal success.

Our team, as in most international organizations, but perhaps more so in the multicultural milieu of Montreal, comprised a group of people with diverse ethnic backgrounds, talents, and personalities. It is inappropriate to only recognize those people in an acknowledgments or a footnote. They deserve recognition in the main text. After all, this is also their story. I will introduce many of them as the story unfolds. There were many others on the team. I have not deliberately omitted anyone. The ones that I recall here are those who mostly influenced me and my own actions.

The story started before I joined Canadian Marconi and I would be remiss if I did not recognize, at the beginning, the early players in our story.

The first product champion was Don Mactaggart, who was brilliant but unconventional in many respects. In the 1970s no Canadian Marconi story or company dinner party would be complete unless at some point one of Don's many escapades was recounted. A Montrealer by birth, Don was the son of a Scottish immigrant. He worked his way through McGill University, earning a degree in electronics engineering, after which he joined Canadian Marconi. Don did not get along with everybody; he was his own person and did not seem to care what others thought of him, although he frequently took newcomers under his wing.

In the middle 1960s Don read about the research of John (Jack) Alvin Pierce at Harvard University. He was working on an idea for a worldwide navigation system that would use a global network of ground stations that transmitted very low frequency signals. Jack Pierce called his idea the Omega System. It was widely expected that the United States would sponsor such a network, but there were no assurances.

Don Mactaggart became convinced there would be a large market for aircraft navigation equipment that used the Omega System, and he assumed the role of product champion. He persuaded Kieth Glegg, then vice president of the Canadian Marconi Avionics Division, to allocate some of the division's precious development funds. He used those funds to start exploratory development of what would be a novel aircraft navigation system. It would also be a much needed new product line for the company.

Kieth Glegg was a visionary as well as a first-class engineer. I consider Kieth to have been one of my important mentors. He was a smallish man with a cheerful disposition, who made everyone feel at ease in his presence. However, his easygoing ways were misleading. If he thought you were wrong or had not properly analyzed all aspects of a suggestion, he could be quite caustic.

I accompanied Kieth on many marketing trips, trying to sell the device that Don and the team had made into a saleable product. Frequently, we carried the heavy equipment to show prospective customers. Kieth, despite his position in the company, was always willing help. He was an eloquent orator, and I learned many presentation skills by watching him at work in front of an audience.

Kieth took a risk with the development of this navigation system and other equally advanced products. The personal cost was high, and sadly when the company's business temporarily declined in the middle of the 1970s, Kieth did not survive the reorganization that followed. Several of the risky product developments he supported, however, did survive and became major revenue sources for the Company.

Kieth and Don did not always agree, and one such confrontation resulted in Don being dismissed. Kieth rehired him some months later. The confrontation occurred after Don had taken the company test aircraft, a twin turboprop Beech model King Air, to South America. The purpose of the trip was to test a prototype of the new Omega Navigation System. As the story, or legend, goes, Don did not exactly have all the requisite company approvals before departure. He finessed his way into several South American countries, got the test data he wanted, and returned with a rather impressive expense account.

Although somewhat overweight, Don had a lot of energy. On a Friday, at the end of a trip to Australia, Don invited the company's Australian representative, who was scheduled to visit Marconi, to his home in Montreal the following Sunday night.

The Australian representative took most of the weekend to travel to Montreal. Don, on the other hand, managed to fit in sightseeing and skiing in New Zealand and a drive around the island of Oahu. Upon arriving back in Montreal, he drove first to a fish market to buy a dozen lobsters and then home to announce to his wife that they were having a dinner party starting within an hour. He had just enough time to put pots on the stove, shower, change, and pour himself a large scotch before he answered the door to greet the Australian representative, who immediately started to complain about the long and tiring trip from Australia and said he hoped that Don had had a more relaxing trip. That, anyway, is the story Don told us that night while we ate lobsters.

Don's escapades would leave us breathless. His penchant for pursuing new product ideas and tirelessly championing them through the concept definition and early design phases made him a true product champion.

Peter Gasser, a Swiss-German with a full dark beard, had recently emigrated to Canada. He was one of the younger engineers on Don's team, who remained with the product long after Don left the group to launch another product idea. Peter, an engineer's engineer, is a practical yet innovative scientist, a leader, and a mentor. Above all, he is a team player.

I worked directly for Peter for nearly a decade as we rose in the company, Peter always one position ahead of me. Don had developed the design concept for the new navigation system, but Peter provided much of the detailed engineering talent. Future generations of the same product were Peter's brainchild as he continued to introduce, or promoted, advanced technologies and innovations.

In this personal account, I cannot mention Peter Gasser without also mentioning Tony Sayegh. Tony did not enter the Canadian Marconi picture until 1982, when he joined our team as a contracts administrator. Tony and his wife, Liliane, emigrated to Canada from war-torn Lebanon. Peter and his wife, Lise, Tony, Liliane, Christina, and I became, and remain, close and trusted friends. Our dinners together are truly international affairs.

Tony's peers soon recognized his talents, and promotions came in rapid succession.

Other members of the pre-1970 Omega product team included Joe Mounayer, another Lebanese immigrant, Ron Miller, an English Canadian, and Michel Galipeau, a French Canadian.

Joe is a remarkable electronics packaging specialist. He has a gut feel for what works, and his design techniques are, at times, rather unorthodox. He does not spend endless hours solving complex mathematical equations but immediately starts building hardware to check out his theories. Because, Joe does not have a university degree, the company has classified him as a technician, not as an engineer.

I tried unsuccessfully on several occasions to have Joe classified as an engineer. I always thought that it was one of the big personnel mistakes we made. I believe he would have felt better rewarded for his years of service if he had been given recognition based on his proven practical track record rather than a 30-year-old educational certificate.

Ron Miller was a solid engineer with an eye for detail. Ron went on to become chief of the Avionics Test Department.

Michel Galipeau was the Omega product manager when I joined Canadian Marconi in 1972. Michel is a gentle man. He motivates those reporting to him without exerting unnecessary pressure. During the years that Michel was my manager, I learned to respect him, and my opinion did not change as his career took him into the world of product quality assurance. In that capacity his sense of fairness and his ability to cope with conflicting obligations to the company and the customer have served him well.

Those people were part of the Marconi team. However, business involves several parties working together. Our customers, government officials, and others outside of Marconi were also major contributors to our success and therefore rightly part of the team. Another early team member was Jack Pierce, for without him there may not have been an Omega System.

The Omega System

Jack Pierce began working as a technician at Harvard University's Cruft Laboratory in 1934. He retired from Harvard in 1974 as a senior research fellow in applied physics. During his 40 years with Harvard, Jack Pierce made many important technical discoveries. Foremost among those was his work in developing navigation techniques that use radio signals with low and very low frequencies.

During the 1950s, Jack Pierce continued to focus on the use of radio waves to improve navigation, with the U.S. military being the chief beneficiary, at least initially, of his work. In 1955, in a classified letter to the U.S. Navy, he reported that a very long range navigational aid, using very low frequency radio signals, was now possible. The Navy did not immediately act on Jack Pierce's proposal, and it languished for several years. Eventually, the U.S. Naval Electronics Laboratory, later renamed the Naval Ocean Systems Center, located in San Diego, conducted some preliminary tests using very low frequency signals from an experimental transmitter located in Hawaii. Those tests were conducted under the leadership of Eric Swanson, who remained a prominent figure in the implementation of the system that resulted from Jack Pierce's work. Eric quickly became an important member of the system's international team, and we worked with Eric in various capacities for more than 30 years.

Jack Pierce's concept did not really take off until the Naval Electronics Laboratory set up a committee in 1963 to produce a design for this new, very low frequency aid to navigation. The committee concluded that a worldwide navigation service could be provided with eight ground stations located approximately equidistant around the world.

The Navy built experimental stations in New York, Hawaii, Trinidad, and Norway. Those stations had only one-tenth of the power considered necessary for the final system, but they were adequate for experimental purposes.

The system that resulted from that early work was called the Omega System. Jack Pierce was asked once, "Who chose the name Omega?" He replied, "I did. In my mind the name meant the far end of the wavelength range useful for radio navigation." Jack understood that it is easier to promote a new product or system if it has a name. He also believed that if a name cannot have a strong and positive denotation, then it should have some interesting connotations.

At Canadian Marconi, we eventually learned that lesson and named several of our Omega products. One was the Arrow-Omega, connoting straight and true navigation. Another was the Alpha-Omega, connoting the first and last in navigation systems.

Many people assume Omega is the name of a company's product or even some secret society. In the early days of launching our product, we would frequently have to explain the background to the name. Customs and immigration officials would quiz us closely when we would try to explain that we were selling Omega Systems. They automatically assumed that we were involved in some clandestine activity. In the world of aviation, the name Omega eventually became well known. Even the U.S. officials who manned the immigration preclearance desks at the Montreal Airport got to know us because we were traveling back and forth so much. They would greet us with, "Hi! Omega boys, off to another fraternity gathering?"

The early experiments proved that the Omega System was a viable concept and that its signals could also be received 30 meters under water, which was just what the U.S. Navy needed for its Polaris submarines. The Polaris submarines used Inertial Navigation Systems, but their positions degrade with time. To compensate for that degradation, the submarines surfaced periodically to allow a crew member to update the inertial system's position by using celestial or other navigational techniques. However, the time the submarines spent on the surface made them vulnerable to enemy detection. Navigation equipment using the Omega System would allow the submarine crew to update the inertial position without surfacing.

In 1968 the U.S. Navy established the Omega Project Office with the funds to complete the construction of a worldwide network of eight transmitting stations. By then the Navy had also contracted with a company called Magnavox to design Omega receivers for ships and submarines. Those receivers, like most other marine Omega equipment, required the operator to tune the receiver manually to the Omega signals. The operator also had to make special measurements and to plot the results on a map to determine the position of the ship.

The U.S. Navy Omega Project Office had the complex task of obtaining the cooperation of other nations where it wanted to locate Omega stations and overseeing their construction. In 1971 the Navy transferred the management of the Omega System to the U.S. Coast Guard. Thus began the transition from a military developmental system to an international civilian and military navigation system.

The six countries originally asked to host an Omega station were Trinidad, Argentina, Japan, Norway, Australia and France. The French station is located on Réunion Island, off the southeastern coast of Africa. The United States planned stations in North Dakota and Hawaii.

The political climate in Trinidad became uncertain, and so the United States looked for another partner. Although on the other side of the Atlantic, Liberia became the preferred location for a transmitter to replace the one in Trinidad. As it turned out, Liberia was also a bad choice. Its proximity to the equator brought unexpected signal transmission problems that we had to fix by changing the receivers. Once built, moving the station was not an option.

Australia proved to be the biggest political problem. New Zealand had already announced it was not prepared to host an Omega station for many of the same reasons that became obstacles in Australia. The problem was that the Omega receivers would be used to help navigate the U.S. nuclear-powered submarines. That led to the perception that the Omega stations would be a prime target during an international conflict. Headlines such as "Australia—Prime Nuclear Bomb Target" began to appear in Australian newspapers. Labor unions, environmental groups, and pacifists staged demonstrations. The Australians' concerns were not based on facts, during World War II, the Korean War, and the Vietnam War, similar navigation stations had been left intact because both sides were using them. Finally, the Australian government agreed to have an Omega station on its territory.

It is probably just as well that back in 1968 the U.S. Congress, the Navy, and the Coast Guard did not know what problems lay ahead. Otherwise they may not have proceeded with what would become one of the navigation bargains of all time, a system used by a worldwide community of land, sea, and air navigators, both civilian and military.

The international agreements and the stations came slowly, one at a time. In 1968 the plan was to have the entire network operational by 1972. The schedule changed almost weekly as political, budgetary and technical problems cropped up. The Omega station in North Dakota was the only station that met the original 1972 schedule. The station in Norway became operational at the end of 1973; those in Hawaii and Japan in 1975; those in Argentina, Réunion, and Liberia in 1976. A full 10 years later than the original schedule, the Omega station in Australia became operational in August 1982.

It would be an understatement to say it was simply a challenge to develop and successfully sell a high-technology product in a sophisticated market in this ever-changing environment.

The Omega stations are large affairs, with antenna wires that are supported by a vertical tower 450 meters tall or a 16,000-meter span across a valley. In Norway the antenna spans a fjord and in Hawaii it spans the crater summit of an extinct, we hope, volcano. Novel construction techniques were adopted in other locations. The umbrella-like station in Japan has its central tower on one island, with the antenna cables stretching from the top of the tower to several other small islands in the vicinity.

The Omega System relies on the principle that very low frequency radio signals travel great distances around the world. For example, a radio signal with a frequency of 10,000 hertz and a power of 10 kilowatts can be heard by a radio receiver up to 10,000 miles away, almost halfway around the world.

Each of the eight Omega stations transmits four primary frequencies, between 10,000 and 14,000 hertz, and other secondary frequencies. The Omega signals appear to a receiver, working with a sophisticated computer program, as a lattice network that, remains in a fixed position around the globe.

The operator of an Omega Navigation System tells the receiver its starting location. Then, as the vehicle in which the receiver is installed travels, the receiver monitors the Omega signals and keeps count of the repeating lines in the lattice network or grid. The Omega receiver compares the signals from several pairs of stations to produce the lines in the lattice network. By measuring the signals, counting the lines crossed since the start of the journey, and calculating where the lines intersect, a receiver can calculate its position on the grid. Although signals from only three Omega stations need to be received to obtain a position fix, Omega receivers process the signals from as many stations as possible, because that improves the quality of the position fix.

Omega signals travel around the world like a tidal wave. The signals are constrained by the Earth's surface and the ionosphere, 70 kilometers above the Earth. Imagine water traveling in a tunnel. If the tunnel narrowed, the water would travel faster as the same volume of water tried to get through the tunnel. The same phenomenon occurs with Omega signals. The ionosphere varies in height above the Earth during a day, with the largest rate of change occurring at sunrise and sunset. Additional variations happen during a year. Further variations, albeit smaller ones, occur during the 11-year solar cycle. Fortunately, thanks to the work of people such as Eric Swanson, most of those variations are predictable, and it is possible to design a computer software program to take them into account.

Accounting for ionosphere variations, termed diurnal corrections, was not the only problem facing the designers of Omega navigation systems. Omega signals are affected by explosions that occur in the sun. The equipment has to know when those events occur and take their effects into account.

Discriminating customers came to expect their Omega receivers to take care of all those variations automatically. At first we thought that the theoretical predictions were completely reliable; we were wrong.

Later, we also found out that Omega signals traveling from a station in Norway to North America were significantly weakened as they traveled across the thick ice of Greenland. That became known as the Greenland effect.

Soon after we installed equipment in aircraft operated by Varig Brazilian Airlines, we noticed that the Omega signals became very distorted in the region of the equator at sunrise and sunset, precisely the time the Varig aircraft, flying between Rio de Janeiro and Lisbon, crossed the equator. The result was that the receiver became confused and large navigation errors resulted.

Omega navigation equipment has an antenna to receive the signals. We found that if the antenna was mounted in the wrong place on the aircraft fuselage, it would pick up false signals, which came from the aircraft electrical system and also confused the receiver.

We had to face those puzzles plus many other technical problems. Fortunately, we had a dedicated engineering team and the will to succeed. As each problem surfaced, we dealt with it. Sometimes the solutions required a significant investment of time and money. Despite all of that, we never reneged on our obligation to our customers. They deserved a product that worked as promised.

Every Omega receiver manufacturer had to face the same problems. Most were generic to the system and not a particular product design. How the companies responded to those problems dictated whether they succeeded or failed. Four companies failed for a variety of reasons. That left three of us to compete for the worldwide commercial airline, business aircraft, and military markets.

Canadian Marconi emerged as one of the two undisputed world leaders in aircraft Omega Navigation Systems. We competed with some of the biggest companies involved in aircraft electronics systems. The company obtained a preeminent position.

Throughout this account you will read about friendships between Marconi employees and individuals working for our customers, our competitors' customers, government officials, and even our competitors. I have highlighted some friendships to show that personal relationships, born out of respect, play a very important role in business, especially successful business.

The Canadian Marconi Omega team, our customers, our representatives outside of Canada, government officials, scientists, and the Canadian Marconi Company management all played their parts as product champions.

This is our story.

2

Pioneering a Product and a Market

Losers visualize the penalties of failure.
Winners visualize the rewards of success.
—Dr. Rob Gilbert

A Product

I can imagine Don Mactaggart sitting at home one evening sometime in the mid-1960s, reading the latest technical papers on the very low frequency navigation technique that Jack Pierce was developing in his laboratory at Harvard. Bored with his work on Doppler equipment, the company's mainstay product line, Don, with his ever-active mind, was thinking about a new product. Perhaps that technique was it. Certainly it fitted his penchant for challenges. An ambitious but as yet unproven concept. A navigation system that would depend upon a network of government-owned and operated ground stations around the world.

If the concept worked, for the first time ever there would be a system capable of providing worldwide navigation to an accuracy of about 2 miles regardless of the length of the trip. The positions plotted by other long-range navigation equipment degraded over time. Jack Pierce also believed that it would be possible to build Omega Navigation Systems for ships and aircraft for half the price of other long-range navigation products.

Don made several trips to meet with Jack Pierce at his laboratory and to fully understand Jack's research, and a kinship developed. At Canadian Marconi, Don began selling the idea for this new product to Kieth Glegg, and with limited funds he produced several design concepts for an airborne Omega receiver.

By 1968 the prospects for the Omega System were very positive. That helped Don finally to convince Kieth Glegg that an Omega Navigation System for airplanes could be a major revenue product for Canadian Marconi. The

Canadian government made Kieth's approval of the project easier by agreeing to share the cost of the development. Kieth put Don Mactaggart in charge of the Omega receiver design, and Don, true to form, began to coax engineers away from other projects.

The system that Don and his team were about to develop had to be as automatic as possible. They knew that pilots would not want to do much more than switch it on, before takeoff, to know the position of the aircraft during the flight.

The marine Omega receiver manufacturers had adopted a different approach. Their systems were low cost, but the operator had to tune the receiver manually, make signal measurements, and then plot the result on a map to know the position of the ship. Such a system was of little interest to aircraft operators.

The marine Omega receiver manufacturers, anxious to capitalize on a potentially large market, also rushed their products to market long before the Omega network was operational. The combination of manual receivers and premature sales resulted in the marine community becoming disenchanted with the Omega System. In the civilian marine world, the Omega System never recovered from those gross errors in marketing. The receiver manufacturers lost a long-term market opportunity for a short-term gain.

The airborne market was a different story altogether. The market matured in a controlled manner. Operators made sure the system worked before they committed to using it, and the government regulatory agencies acted conservatively. Those of us trying to peddle the products as if our careers depended upon them occasionally wished that the market was a little more naive.

Companies that are involved with a product for which they do not have complete control over all aspects of the development must understand that they are taking some considerable risk. Even when they have control, they still face the vagaries of the marketplace. For some, the market risk alone is risk enough. Only brave companies commit their resources to product development when they do not have at least some initial control. Companies developing products and systems that rely on a government-sponsored supporting infrastructure, like the Omega System, take the ultimate risk.

I have come to realize that government employees, particularly those involved in technological projects, are competent and dedicated. What government programs suffer from is changing political and fiscal priorities. Those, coupled with the usual technical problems that crop up, lead many to shy away from government-sponsored programs. For those that understand the rules, if there are any, the rewards can be worthwhile.

There is no magic formula. Prerequisites are a thorough knowledge of the project, participation in the system's evolution, close ties with the government agency involved, and experience in that market. Timing and knowledge are the keys to success. A little luck also helps.

Toward the end of the 1960s the Canadian Marconi Company's Avionics Division needed to broaden its product base to grow or even to survive. The

Doppler Navigation Systems continued to be a commercial success, but it was risky to be a one product division. The company had succeeded in selling a few Doppler systems to airlines, but its success in the worldwide military market had been spectacular.

During the late 1960s the airlines became interested in a new inertial technology for navigation that first had been developed for guiding missiles and, later, military aircraft. Unlike Doppler systems, the inertial equipment did not send out radio signals. It also did not rely on signals from ground stations as did many other navigation systems in use at the time. The inertial products contained spinning gyroscopes that tilted when the aircraft accelerated. An Inertial Navigation System measures the displacement of the gyroscope and combines that with other information to calculate the aircraft's position.

The Boeing Commercial Airplane Company's B-747 jumbo jets were the first commercial airplanes to use Inertial Navigation Systems as the principal means of navigation. They were expensive, but they were a formidable competitor to the much less expensive Omega Navigation Systems that Canadian Marconi and others were about to develop.

For many years the civilian community was skeptical about the Omega System because the military was the system's key sponsor. One of our biggest challenges was convincing the airlines that Omega was not exclusively a military system.

That was the environment that Canadian Marconi faced when it launched the development of an airborne Omega Navigation System. It was a risky venture.

Companies do not develop complex electronic products overnight. Their engineers have to assess and to test new technologies. They have to try various design concepts and to build prototypes. Once the engineers are confident about the basic design, work can begin on the detailed design of a product. All that takes time.

A particular challenge for avionics designers is squeezing all the electronics into as small a box as possible. They have to keep the weight of the product to a minimum and to produce a design that consumes the least amount of power, a resource that is scarce in an aircraft.

Aircraft shake and rattle, sometimes have hard landings that jar the systems on board, and operate in extreme temperatures. That means that the engineers also have to design avionics to withstand a rugged environment.

Kieth Glegg, with his infectious laugh, would often say, "Work should be fun; if it isn't, then you are in the wrong job." He was convinced that workers who are having fun work harder and with more dedication than those who are not enjoying their work. Hard work and dedication can be the difference between success and failure.

In 1968 the Canadian Marconi engineers on Don's team were very motivated as they began the development of what would become one of the world's first airborne Omega Navigation Systems.

In the United States the large Northrop Corporation was also developing a system. The company had won a development contract from the U.S. Navy. It had a head start over Canadian Marconi and, more important, a funded development contract. Canadian Marconi's project was a private risk venture with some financial help from the Canadian government.

An aircraft Omega Navigation System has three parts: an antenna that is mounted on the outside of the aircraft to receive the very low frequency signals transmitted from the Omega ground stations; the black, sometimes battleship grey, receiver-computer unit that contains most of the system's electronic circuits, is the heart of the system, and is installed in an aircraft compartment that houses all the electronics boxes that are the brains of any aircraft; and a small control and display unit. The pilot does not see, or care to see, the antenna or receiver-computer unit. His connection with the system is the control and display unit mounted within easy reach in the cockpit.

The front face of the early control units had a rotary switch for selecting the navigation parameters that were available from the system. Those included the aircraft's position in latitude and longitude, the distance to go to a location farther along the flight path, and the speed of the aircraft. The system could calculate whether the aircraft was on course and, if not, how far and in what direction it was off the intended flight path. The pilot used a keyboard, like that on a calculator, to enter data such as the latitude and longitude of the takeoff position. Above the rotary switch and keyboard, a digital readout displayed the navigation information.

Peter Gasser led a small team of engineers in the design of the Omega signal receiver. Joe Mounayer, the electronics packaging specialist, worked first on the receiver-computer unit and then, with Ron Miller, on the control and display unit. Jean-Claude Lanoue, a recent immigrant to Canada from France, and Michel Lavoie and Gilbert Boileau, both French Canadians, handled the software.

Michel, dark haired with boyish facial features, Jean-Claude, with a more serious disposition, and the quiet hard-working Gilbert were the backbone of the Omega software team for many years. I learned a lot about the Quebecois from Michel and respected, but did not agree with, all his views on Quebec's place in or outside the Canadian Confederation. However, our political differences did not prevent us from becoming close business colleagues. Michel also helped me when I was struggling to learn and mispronouncing French.

Jean-Claude's disposition did not endear him to everyone, but those who took the time to listen to him learned a great deal. I never pretended to be a design engineer. Perhaps because of that I was a little less inhibited than others about asking Jean-Claude to explain, sometimes more than once, a technical matter. Later Jean-Claude moved from software to hardware design. More than anyone else, he introduced world-leading technologies and innovations into our Omega Systems designs.

Elsewhere in the Avionics Division, David Woodhouse was working on a new avionics computer. The Woodhouse computer, as we affectionately called it, found its way into three major new products, one of those being our first Omega Navigation System.

Initially, Don Mactaggart led the design effort, but his impatience for results frustrated many of the engineers. Such was the price to pay for being a true product champion. Michel Galipeau, the gentle man, was brought in as product manager to keep peace within the work force. With that change Don had more time to help solve some of the unique design problems that the engineers were encountering. He also kept in touch with Omega events unfolding outside the company.

The Canadian Marconi Omega System design effort continued unabated for two years. The usual plethora of problems arose, and most were solved. By 1970 the engineering team had completed and tested on the bench a prototype of the product. Then they carried out crude, but effective, mobile tests up and down the local freeway with the equipment laid out in the back seat of the project engineer's car. Finally, the equipment was ready for the big test. It was time to go flying. A series of flight tests were carried out successfully in the company's test aircraft, but it was not enough.

It was preferable for Canadian Marconi to ask for the support of a major organization to test and, therefore, indirectly to help with the completion of a product's design. In those days that meant seeking the support of a military organization. The Canadian Armed Forces had expressed only a lukewarm interest in the Omega System and the U.S. Navy was already helping Northrop. Canadian Marconi had to look elsewhere.

With his penchant for foreign travel, Don had maintained close contact with Britain's Ministry of Defence, Royal Aircraft Establishment, at Farnborough, and with good reason because the British had a strong interest in Omega. In fact, they were carrying out experimental Omega work with a British company called Redifon. With the Canadian Armed Forces and the U.S. Navy ruled out, the British Ministry of Defence became a logical next choice.

The Marconi folks knew that a flight test report issued by the Royal Aircraft Experimental Establishment would help their Omega marketing effort. Britain's Ministry of Defence, especially the Royal Aircraft Establishment, was known and admired throughout the world. Such a flight test program also would provide valuable information for the Marconi engineers, who were trying to complete the product.

The Royal Aircraft Establishment agreed to fly the prototype system in its Comet test aircraft. The first series of flights took place between October and December of 1970.

The Omega System that was flown did not look much like the final product. It was a typical prototype system. The computer, an early version of the one that David Woodhouse was designing, was in its own box. The Omega receiver was in a makeshift box, and the control and display unit was a very crude version of the final product. The units were bulky and connected with a mass of wires. The Omega installation also included a recorder and various pieces of test equipment.

Peter Gasser, Don Mactaggart, John Boss, and Colin Gyles set off for Farnborough with crates of equipment, tools, equipment circuit diagrams, and spare electronic components. Within a few days the team, working with the aircraft technicians, completed the Omega installation in the Comet aircraft.

At the time of the flight tests, only four experimental Omega stations—Norway, Trinidad, New York, and Hawaii—were on the air. Those stations were transmitting at only one-tenth of the power planned for the final network. Worse, they only transmitted on one frequency and not four as eventually planned. The station on the Hawaiian island of Oahu was too far away. There was no trace of its solitary signal, and so only three stations were available for the flight tests. The receiver was going to have to perform against considerable odds. However, the tests proceeded.

That situation was normal prototype product testing in an experimental environment. However, Canadian Marconi management was anxious for a sale. The company did not view those tests as being purely experimental; it hoped that the tests would lead to an order from the British government within a year or so. After all, the U.S. Navy and U.S. Coast Guard expected to complete the construction of the eight-station Omega network within two years.

Everyone was overly optimistic. It was not until 1976 that seven of the final Omega stations came on the air, six years after of the Farnborough tests, not two years.

The Royal Aircraft Experimental Establishment flew Canadian Marconi's prototype Omega Navigation System in its Comet on two routes. One was between Farnborough in the south of England and a point some 150 miles north of Scotland. The other route flew over northern Ireland. The results were surprisingly good; the system navigated to an accuracy of 2.5 nautical miles most of the time. The results were also within the U.S. Navy's predicated accuracy for the Omega System.

The report of the flight tests, written by Colin Gyles and dated Dec. 15, 1970, stated, in part, the following conclusions: "The biggest worry with Omega receivers is lane slippage. The results of these 7 flights show very encouraging results with only 1 lane slip out of a total of approximately 550 lanes traversed, and that when flying through very high precipitation static. The static will be greatly reduced when a loop antenna is used in the next series of flight trials."

The report's conclusions were cause for optimism. However, the two primary observations were portents of trouble and much engineering.

The reference to lane slips seemed innocuous, and the results were even encouraging. An Omega lane is the distance between the lines in the Omega signal grid. If an Omega receiver miscounts those lanes, called a lane slip, navigation errors occur. The principal causes of lane slips are electrical interference or disturbances in the ionosphere. We had to invest considerable resources to overcome those disturbances. Some companies did not find solutions or were unwilling to make the investment necessary, with the result that their products failed.

The conclusions to the report also referred to precipitation static. For the Comet tests, the Omega antenna was a wire outside the aircraft that picked up the electric field component in the Omega signals. The problem is that static electricity builds up on the skin of an aircraft, particularly in thunderstorms, and interferes with the Omega signal. The interference confuses the receiver, which, in turn, causes navigation errors. To overcome that problem Omega receiver manufacturers designed a special Omega antenna that picked up only the magnetic field in the Omega signals. That special antenna is the loop antenna referred to in the conclusions to Colin Gyles' report.

Rather belatedly, Omega System engineers found that an aircraft power system could interfere with the signals being picked up by the loop antenna. By a quirk of circumstance, the frequencies that Jack Pierce had chosen for the Omega System matched the harmonic frequencies of the standard power supply on all aircraft.

To overcome that problem, it became necessary to find a location on the aircraft skin where that interference did not exist. Often there was only one such location on the aircraft. If a suitable location could not be found, an electrical field antenna had to be used and the static electricity interference accepted.

We called the process of finding a suitable location for a loop antenna on the aircraft skin mapping. To perform a skin map, our field engineers had to walk around an aircraft with a special noise detection instrument. It was precarious work because they had to do that job when the aircraft engines were running. Because the loop antenna was far superior to the electric field antenna, we went to considerable lengths to find a suitable, noise-free location.

No one said it was going to be easy to launch the Omega System. We just hoped that it would be. As one problem surfaced and was solved, we convinced ourselves that was the last one. This was the only way we could maintain our determination to succeed. However, for years there was always one more problem.

The results of the Comet tests provided the impetus that the Marconi engineers needed to press ahead with completing the design of the product. Don Mactaggart became even more impatient. Late nights and weekend work became the order of the day. Although the ground stations were still in an experimental status, there was optimism that the network would be fully operational as planned in 1972 or very soon thereafter. The Omega concept held the promise for a breakthrough in navigation, and the market seemed limitless, provided the ground stations came on line on time.

Tony Lee, Canadian Marconi's Omega applications and marketing engineer, was particularly optimistic about the market potential. In several marketing analyses he suggested that Marconi should immediately begin building a separate production facility for the Omega product line. Fortunately, Canadian Marconi management did not share Tony's optimism because the market took longer to develop than Tony's optimism suggested.

The team was coalescing and so was the product. Half of the team continued to work on the prototype system, finding solutions to the variety of design problems that surfaced as testing against real signals took place. The other half of the team worked on refining the electronics design to fit into an avionics package.

Because the Inertial Navigation Systems would be the prime competitor to Omega Navigation Systems, Canadian Marconi engineers decided to make their Omega System appear, on the outside, to be the same including its aircraft wiring. That included providing most of the same navigation features available from the Inertial Navigation Systems. They expected the strategy to help to convince prospective Inertial Navigation System customers to change their minds and to buy Omega Navigation Systems. At least a customer would not have to make drastic changes to an aircraft that had already been wired for an Inertial Navigation System.

It was a good strategy in theory, but it did not exactly work out that way in practice, even though the Omega Navigation System was half the cost of an inertial system. One problem was that the company had underestimated the time it would take to convince the market and the regulatory agencies that the Omega System was viable, safe, and accurate.

The first Canadian Marconi airborne Omega Navigation System was assigned the part number CMA-719, not a very catchy name. The CMA-719 receiver-computer unit was about the size of a filing cabinet drawer. The antenna measured 12 by 12 by 3 inches thick, and the control and display unit was similar in size to two car radios stacked one on top of the other. The whole system weighed 50 pounds.

One of my challenges in marketing that product was lugging the heavy and bulky equipment around the United States, Europe, and South America. We packed the equipment in boxes sturdy enough to survive being thrown around by airline baggage handlers. A suitcase contained power supplies to run the equipment in conference rooms and a slide projector was safed in a metal case. The Omega System traveling sales kit consisted of four lugable boxes weighing over 100 pounds.

I was a very happy salesman when the next-generation equipment came out at one third of the weight of its predecessor. Fifteen years later, the fourth generation system weighed less than 10 pounds.

One of the true wonders of the first equipment was the memory device used to store the computer software. Compared with today's technology the computer storage device was positively archaic. The CMA-719 memory, like those of other computers of the day, had a magnetically energized ferrite core.

Electrical signals charged the individual magnetic elements in either a positive or negative fashion to represent the ones and zeros upon which the binary computer bases all calculations. Considering its electromagnetic characteristics, the ferrite core memory was very effective and usually reliable.

On some occasions, though, we would hear an engineer or an irate customer exclaim, "The system has blown its brains again." In computer jargon the term "the system has crashed" is used more often. It was most frustrating to be in the middle of an important demonstration and have the system crash. It was even more disturbing for some of our early customers, who experienced system crashes at the most inopportune times. We gave one customer, who was operating in remote regions of the Arctic, their own memory loading device, because their systems were crashing so often. Fifteen years later, the fourth-generation system had a memory 60 times larger, in one-tenth of the space, and was 100 percent reliable.

The computer electronics of our first system were contained on 10 electronic circuit boards in the receiver-computer unit. The fourth-generation Marconi Omega System had a computer eight times faster mounted on only one board. Similarly, the first 3-frequency receiver occupied 10 circuit cards, and later a receiver capable of receiving 12 different frequencies occupied just 1 card.

Such was the technology revolution we would experience over the years. In the meantime, we had to find a market for our first-generation system.

The Search for a Customer

By the summer of 1971 the CMA-719 was beginning to look like a system that could be routinely installed in aircraft. Gone were the myriad of individual units and mess of wires. In their place were the three units that made up the system, all packaged to withstand the rigors of the aircraft environment. To accommodate both military and commercial aircraft installations, the Marconi engineers decided to produce two versions of the CMA-719. Both contained the same electronics but were packaged differently.

The CMA-719M receiver-computer circuit boards were housed in a rugged case to protect the electronics from the extreme environment in which a military aircraft has to operate. The CMA-719C, for commercial aircraft, was less rugged, lighter, and less expensive than the military version. The engineers spent most of their energies finishing the two designs and readying them for production. Tony Lee, Marconi's Omega applications and marketing engineer, persuaded the Canadian Ministry of Transport to test the system in its Douglas Aircraft Company DC-3. The ministry used that aircraft to monitor ice movement in the Arctic as an aid to shipping.

In September, Britain's Royal Aircraft Establishment once again tested the CMA-719 in its Comet aircraft. The system was flown throughout Europe where it navigated on average to within 1.5 nautical miles.

The product was finally ready for production, but where were the customers? Tony Lee tried to convince the U.S. Air Force to begin procurement before Marconi had so many orders that it would have to wait in line—wishful thinking. Tony's frequent trips to Wright–Patterson Air Force Base, Ohio, were to no immediate avail. The Air Force scientists were extremely interested in the pioneering Omega work that Canadian Marconi was doing, but the Air Force did not have a requirement for Omega Systems, and the Congress had not approved funding to buy systems.

The U.S. Navy, which had a requirement, also had funded the Northrop Corporation to design an airborne Omega Navigation System.

Undaunted, Tony Lee frequently visited the U.S. Navy airborne equipment project office, called the Naval Air Systems Command, in Washington, D.C. He repeatedly tried to persuade the Navy that the Canadian Marconi equipment was better than Northrop's. Over the years, I would also spend much time on the same impossible mission.

Eventually, we learned another in a long list of lessons. The most difficult market to penetrate is one in which a competitor's product is already established, even when your product is better. Buyers are reluctant to admit they may have made a mistake in selecting the first product, and they are not anxious to repeat the bidding process. In the situation we faced with the U.S. Navy, we knew our product was not appreciably better than Northrop's. We even suspected that Northrop's software was better than ours. We did not have a strong case. With the Navy, we were doomed to fail before we even started. We just did not know it or did not want to know—at the time.

By 1971 the U.S. Navy had tested the Northrop Omega System in several different aircraft including helicopters. The system had also been granted the all important NATO nomenclature. Once a NATO country buys, or in some cases tests, a piece of avionics they register that product with a special military number. In the world of military marketing, the product has more credibility if it has a NATO number. A NATO number suggests that the product has some pedigree.

It did not seem likely that either the U.S. Air Force or the U.S. Navy were going to be early customers of the CMA-719 so Canadian Marconi had to look elsewhere for a launch customer.

While maintaining regular contact with the U.S. Navy and U.S. Air Force, Marconi turned its marketing effort toward Europe and Great Britain's Royal Air Force. To help the effort, Canadian Marconi signed a collaborative Omega marketing agreement with Marconi Avionics in the United Kingdom. The arrangement effectively split the world in two, but Canadian Marconi would manufacture the Omega Systems required for both companies.

The arrangement, which was sound in theory, did not work out as well as expected in practice. It was too ambitious, and the territories were too large for such a new product with so many problems still to be solved. Also, Marconi

Avionics, like Canadian Marconi, is a manufacturing operation, and it is difficult for people in a manufacturing business to be motivated enough to succeed when selling someone else's product just for profit. Such sales do not keep the factory active. Engineers and marketing professionals prefer to work on and sell their own products.

Eventually we revised the joint marketing arrangement so that each sale could be treated on a case-by-case basis. In this way we could take advantage of either company's strengths in a particular market or with a particular customer. The less ambitious arrangement worked well, and the two companies cooperated on several Omega sales. Marconi Avionics also provided invaluable Omega System service support to our European customers over the years.

In the beginning both parties held great expectations for the arrangement. To test the CMA-719 and to demonstrate it to prospective customers, Marconi Avionics installed a preproduction version of the equipment in its test aircraft. Their Piaggio test aircraft, with its two rear-facing propellers, could be heard miles away as it droned over the Essex countryside. Earlier, as a Marconi apprentice, I had spent many hours testing a new Doppler Navigation System in the same aircraft.

Sterling Airways, a small commercial charter airline in England, was the very first airline to show interest in the Omega System. It asked to borrow a CMA-719 system to test in its Caravelle aircraft. Canadian Marconi readily agreed and scheduled the test to start in September 1972.

While all that pioneering Omega work had been going on, I had been working for Marconi Avionics U.K. as its representative at the Boeing Company in Seattle. I had only a superficial knowledge of Canadian Marconi's Omega program, and most of what I did know was due to Boeing's interest in the system.

At that time Boeing tried to enter the maritime patrol aircraft business, a market dominated by the Lockheed Aircraft Company with its turbopropeller P-3 aircraft. Boeing decided to offer a jet version using the B-707 as the basic airframe. The U.S. Navy already had a large inventory of P-3 aircraft. However, the Canadian Armed Forces announced that it would soon be replacing its aging Argus maritime patrol aircraft, becoming a possible launch customer for Boeing. To make the airplane as enticing as possible, Boeing promoted Canadian equipment for the airplane, including the Canadian Marconi Omega System. Marconi loaned Boeing a CMA-719 system to use in its avionics integration laboratory and later in the test and demonstration aircraft.

The procurement cycle was typical for a military program, and it was not until several years later that the Canadian government made a final decision. In the meantime, the CMA-719 remained at Boeing, along with other Canadian Marconi equipment that was also on loan, and Marconi routinely sent engineers to Boeing to help it with its work.

In the end, the Canadian Armed Forces chose a modernized version of the maritime patrol workhorse, the P-3. In 1972, however, both Boeing and Canadian Marconi saw the program as a possible launch for two new products. The early part of the program also gave me an opportunity to become acquainted with the Omega System and, more importantly for me, the Canadian Marconi Company.

Three years in Seattle had whetted my appetite for North America, and during my last six months there I had been trying to secure a position at Canadian Marconi. By July 1972, I had completed my planned tour of duty at Boeing, and because I still had no offer of employment from Canadian Marconi, I prepared to return to England.

3

I Join the Canadian Marconi Company

I am looking for a lot of people who have the infinite capacity to not know what can't be done.
—Henry Ford

Three years as a company representative at the Boeing Commercial Airplane Company had passed all too quickly. During the summer of 1972 Christina and I left Seattle. A few weeks earlier, during a visit by my parents, we had made one last camper van tour of the Pacific Northwest. It was a splendid way to say farewell to a beautiful part of North America after three wonderful years of enjoying all that the region had to offer. I had become an active scuba diver. We had skied and hiked in the Cascade Mountains. Twelve months earlier, a friend had introduced me to ice climbing with an ascent of Mount Saint Helens—before it erupted. And I had met my future wife, Christina, a tall, slim redhead, in 1970 while skiing north of Vancouver.

I returned to England, via Montreal and India. Marconi business caused my circuitous route to England. Marconi Avionics and Canadian Marconi had teamed up to pursue a business opportunity with the Indian Air Force. The marketing manager in England asked me to go first to Montreal to help put together our presentations and demonstration equipment. Two weeks later, in India, Bill Buker, of Canadian Marconi, and I joined up with my Marconi Avionics colleagues. This was my first visit to India, and from a business perspective it was a most frustrating experience. We spent many hours in airless rooms awaiting the pleasure of various Indian Air Force officers.

Christina was excited when early in 1972 I began to try to secure a job at Canadian Marconi. In 1962 she had emigrated from England and subsequently had lived in Montreal before moving to Vancouver. She was anxious to return to Montreal. Upon my return to England, the Marconi Avionics marketing department offered me a job, a position that I readily accepted because a career in avionics marketing had been my goal ever since completing the Marconi engineering apprenticeship some six years earlier.

Finally, on Aug. 3, 1972, the correspondence that I had waited nine months to receive arrived in the form of a telex to the marketing manager.

> ATTN: JIM COURT
> PLEASE ADVISE SITUATION OF G.GIBBS STOP WE ARE IN NEED OF MARKETING PEOPLE AND WOULD LIKE TO CONSIDER HIM IF HE IS STILL DETERMINED TO RETURN HERE STOP
> REGARDS ROSEBERY.

Jim Court was not anxious to see me go. However, he did admit that better career opportunities probably awaited me in North America.

After the first telex, events moved quickly. There was an exchange of letters, and by Aug. 25, Canadian Marconi offered me a job at a salary of $11,000 CDN per year, an offer that I readily accepted. I was to be an applications engineer in the Avionics Division.

The Avionics Division uses the title applications engineer to indicate, to customers that the individual is knowledgeable in the design, installation, and operation of the products that he or she is marketing. The practice of using only technically qualified personnel in marketing positions has served the company well over the years. No customer, especially one in a high-technology business, wants to deal with a Hollywood version of a salesman.

Originally, I was to be the applications engineer for a new navigation management system that the division was developing. Luckily for me, that did not come about as a result of the unexpected resignation of Tony Lee, the Omega applications and marketing engineer.

While I was at the Farnborough Air Show, at the end of August, working at the Marconi Avionics display booth, another telex arrived.

> ATTN: J. COURT
> FURTHER TO MY MESSAGE CONCERNING G GIBBS IT WOULD BE DESIRABLE FOR GIBBS TO ARRIVE HERE SEPT 12 OR 13 TO OVERLAP WITH LEE WHO IS LEAVING ON SEPT 15 STOP
> REGARDS ROSEBERY.

They wanted me there within two weeks. Worse, I was only going to get two days of overlap with Tony Lee. That was not the first time, and it would not be the last, when I would have to take on a new task and survive the best I could.

That same day, at the Farnborough Air Show, I met Don Mactaggart for the first time. I literally bumped into him, or rather he bumped into me. As usual, Don was dashing somewhere with his mind on some knotty technical problem and not the world before his eyes. I introduced myself. He was almost curt and said very seriously, "So you are joining Canadian Marconi? Well, you've got an easy job. The Omega System will sell itself."

The Canadian Marconi Omega team members were still in their optimistic frame of mind. Tony Lee was still telling the management that it should heed his market projections and begin building a new factory to cope with the expected large production of Omega Systems. The full impact of the delay in the Omega ground stations had not yet sunk in. Nor had they understood how cautious the aircraft operators and the regulatory agencies would be to embrace a generically new system without a lot more testing.

After some hasty telephone calls to Jim Court, I left the Farnborough Air Show. My first task was to break the sad news to my parents and the good news to Christina. We did not have much time to take care of the affairs we had to settle to be ready to emigrate to Canada. Also, the job offer had come with little time to spare, because although I had a visa for Canada, it was about to expire.

My parents, to their credit, supported my sister and me whenever we wanted to do something that would improve our lives. They had put on a brave face when my sister emigrated to New Zealand, and now they prepared themselves to cope with my departure for Canada. Some years later, when my farther retired, my parents followed my sister to New Zealand. The temperate climate of New Zealand won over the frigid winters in Montreal and the wet climate of England.

On Sept. 12, 1972, Christina and I left from London's Heathrow Airport for a new adventure, a new life, and, for me, a challenging and exciting career. We arrived early on a Wednesday morning. Tony Lee planned to leave the following Friday, and so I decided to go straight to the Canadian Marconi plant. Christina went to the house of friends.

Christina was the wife, mother and homemaker who took care of most household matters. She was the parent who was always there to care for the children and help solve their daily problems. That support allowed me to cope with a frenetic work and travel schedule during my years with the Canadian Marconi Company. No matter where I was in the world I always telephoned home regularly to talk to Christina and our girls. At times, I felt as if I were carrying out my parenting only over telephone lines. We all understood that anyone in the family could call me at any time, no matter where I was, or what I was doing. That was the only way that we could maintain a semblance of family communication, when I was in a hotel more often than at home.

Within two weeks, Christina had us moved into an apartment, which was in the fashionable community of the Town of Mount Royal close to the center of Montreal. It was also just a short walk to the Canadian Marconi plant. We bought a bed and that, together with a borrowed portable games table and two chairs, constituted all of our furniture. In the following weeks, we made our own kitchen table, two couches, and other pieces of furniture. We had some savings, but after buying a second-hand car, we wanted to start saving for a house of our own. We were a typical young couple, poor as church mice. Christina started working at a bank, and I settled down at Marconi.

At Canadian Marconi, on that first Wednesday, I was feeling the effects of the overnight flight that we had just made. Sleep or no sleep, I needed to spend as much time as possible with Tony Lee before he left. We agreed to postpone the usual visit and form filling with the personnel department. I had two and a half days to extract from Tony all his knowledge about the Omega System, the Marconi product, and the market opportunities that he was pursuing. It was a difficult task, but I need not have worried because I was joining a team.

The team had its own territory, which we called the Omega Lab. Within the four walls were the product's design engineers, technical managers, contracts administration staff, draftsmen, and engineers. We were and remained a self-contained unit with everyone readily to hand. The model shop, where craftsmen could quickly build prototype equipment, was next door to the Omega Lab. The production facilities, located one floor below, were the only key element not also located with the Omega Lab. Even representatives from the Field Engineering and Training departments were collocated with the Omega team.

It was a convenient arrangement. Combining all personnel associated with a product line into one area was a unique feature of Kieth Glegg's product management matrix organization. In that way, the product manager and the entire team had quick access to each other. Locating a product team together and integrating all engineering functions was, I believe, a key element in our ability to remain competitive.

I met the team, most of whom I have already introduced. Michel Galipeau was the product manager. Don Mactaggart, true to character, was now pursuing another product idea, a navigation system for ships using signals from a constellation of U.S. Navy satellites. Peter Gasser was the project engineer. Jean-Claude Lanoue, Michel Lavoie, and Gilbert Boileau were developing the complex Omega software. Ron Miller was working on the Omega System's Control and Display Unit, the essential link between the system and the pilot. Joe Mounayer, the electronics packaging specialist, was finding ways to cram all of the electronics into standard-sized aircraft avionics boxes. David Bailey, a quiet-spoken English Canadian and recent graduate, was also on the team with about eight other engineers, technicians, and draftsmen.

Many of those people now have senior positions within the company. Their involvement in the Omega product line and their skill and dedication in searching for and finding excellence helped their careers.

The first half week that I spent at Canadian Marconi passed all too quickly. Tony had given me the 100-paged technical description of the CMA-719 Omega Navigation System, which the team had written for prospective customers, and a pile of other documents. For several weeks those were my only bedtime reading.

Tony Lee had mostly pursued military business opportunities with the U.S. Navy and Air Force, the Canadian Armed Forces, and the British Royal Air

Force. He had also ventured into the commercial aviation arena, developing a few marketing leads with operators such as Sterling Airways of England and Nordair of Montreal.

Tony still felt that the real business opportunity for Canadian Marconi was with the U.S. Air Force. He gave me the name of a particular contact at Wright–Patterson Air Force Base, who he thought was our best ally. Wright–Patterson Air Force Base, in Dayton, Ohio, is named after the aviation pioneers. There, the U.S. Air Force carries out much of its aircraft systems research, as well as most of the testing and procurement of systems required by an Air Force operations command. Tony told me that the engineers and scientists at Wright–Patterson Air Force Base were very interested in the Omega System, although he suspected that interest might be a result of their jealousy of the Navy's lead as the Omega System managers. He guessed they were also jealous of the Navy's Omega receiver development program with the Northrop Corporation. The scientists at Wright–Patterson felt left out; they wanted their own Omega program.

The problem was that no Air Force commands, the real customers, had identified a requirement for airborne Omega Navigation Systems. On the contrary, they were hoping to continue installing the much more expensive Inertial Navigation Systems. That was mainly because the inertial systems did not rely on signals from ground stations, signals that, in theory, an enemy could electronically jam.

During frequent visits to Wright–Patterson, Tony and others kept hearing that the Air Force commands would soon begin to appreciate the benefits of the Omega System, which would lead to the purchase of thousands of systems. If their marketing had covered the Pentagon and the Air Force commands with as much fervor as Wright–Patterson Air Force Base, they might have heard a different story. The operators would have clearly told them that they had no requirement for Omega equipment. Then the company would not have been so optimistic about the near-term market potential. Like a journalist, a marketeer must confirm information with several sources before using that information as the basis for a market analysis, especially an analysis that will influence management decisions that may prove expensive for a company.

That is the negative side of the story; the positive side is that any company selling a new product must remain optimistic, while being objective. Customers are secured through marketing, communicating the idea, building products that respond to requirements, and selling the advantages of the new product. Tony and the Marconi Omega team were doing all of those tasks. They were pioneers in the marketing of a new concept in navigation, and there was no shortcut to success. Their only weakness was that they were too optimistic about the near-term market potential. It was a weakness that we all shared as we continued to believe that our elusive customers were just months, not years, away. Our optimism kept us going.

Tony left and I was a marketing department of one, at least on paper. However, we were a team.

In December 1972, Nordair, a Montreal-based operator, became our first customer. It had recently secured a contract with the Canadian government to provide year-round patrol of the Arctic to monitor and to report on ice movements. It planned to use two Lockheed Electra aircraft especially modified for the task.

Tony Lee had convinced Nordair and the Canadian government that they should equip the Electra aircraft with Omega Systems, to back up the Inertial Navigation Systems that were the principal source of navigation information.

That first order was for two systems and was valued at about $100,000. It was a start but was far short of being the large launch order that we needed.

A year earlier, Kieth Glegg had authorized a pilot production run of 10 CMA-719 systems. Those were the rugged militarized version. Now we tried to convince Kieth to authorize the production of 100 because we felt that the market was about to take off. But every time we thought that was going to happen, the Omega ground station schedule would slip again. Kieth was not ready to take the risk of investing the large sum we would need to build 100 systems.

All was not well with the system design either, and so there was still much engineering work going on. The principal problem facing the engineers was the lack of reliability of the ferrite core memory, the device that contained all of the system's software. Without warning or provocation, the memory would become unstable, and the software would scramble. That meant that the entire software program had to be loaded back into the memory using a digital tape player and special test equipment.

We had only built two such pieces of equipment, one for the Production Department and one for the engineers. Sometimes the engineers needed both sets to keep the laboratory equipment running because the memories were scrambling so often. At other times production needed two setups. One result was that the engineers and the shop floor people got fit running between production on the second floor and the Omega Lab on the third floor.

In the autumn of 1972, when Sterling Airways began its flight tests of the CMA-719, the memory problem became complicated. Every time the system lost its mind, we had to send a field engineer to England to reload the software. Eventually, we loaned Marconi Avionics, which was supporting the evaluation, one of our two pieces of software loading equipment, but that simply irritated the problems back at home.

The first time that the Nordair systems lost their memory, we reloaded the software and kept our fingers crossed. Two weeks later a system scrambled again. We sent an engineer to Frobisher Bay, on Baffin Island, to intercept the aircraft and reload the software. We went on like that for months. Meanwhile, the engineers feverishly tried to find out what was causing the memories to crash and quickly built more software loading equipment.

Our other major problem was the one described by Colin Gyles in his report on the first flight tests carried out by the Royal Aircraft Establishment. Occasionally, after the system had been navigating accurately for some time, it would slowly, or sometimes abruptly, drift off course. The slow changes were the most troublesome. Those could go undetected unless the pilot could use another system on board the aircraft to cross-check the Omega equipment position.

As the navigation error increased, the system would lose count of the Omega signal lanes. Then the only solution was for the pilot to update the position using information from another navigation system—if one existed.

Those random navigation errors did not help us as we tried to convince the world that an Omega Navigation System could be the primary or only means of navigation on an aircraft. The lane slips were probably the most serious of the two major problems and the myriad of lesser ones that our engineers were trying to resolve. If we could not solve those problems, no regulatory agency was going to certify the equipment, and we would not have any customers.

Other problems that we had to cope with included disturbances in the ionosphere, which were affecting the Omega signals. Also, discrepancies between what we thought the ionosphere should be doing at a certain time and what was actually happening were affecting the accuracy of the system. Our basic ionosphere variation model obviously had some errors.

U.S. government scientists had produced a comprehensive model of the behavior of the ionosphere, but it contained too many algorithms to be stored in the early avionics computers. Several years later airborne computer technology had advanced so much that we could store the complete variation model. In the meantime, we used an abbreviated version that Eric Swanson, of the U.S. Naval Electronics Laboratory in San Diego, had developed. Unlike the complete prediction program, this version could be contained in an airborne computer. Canadian Marconi and others were using Eric's model, and throughout the industry we called it the Swanson Model. At the time neither Eric nor our engineers had any reason to doubt the accuracy of the predictions upon which Eric had based his model.

As it turned out, Eric's model was flawless. The problem was a combination of the accuracy of the basic prediction model and how the receiver manufacturers were implementing it in their equipment. Several years passed before we fully understood the behavior of the Omega signals and the ionosphere.

As the engineers solved one problem, another one would crop up. The seasons were changing, and so was the variation of the ionosphere, but the variations were not as predicted. The discrepancies in the seasonal variations were not the only problem. The system would work in one location and everyone thought we had finally solved the problem. Then a prospective customer would fly the system in another region of the world, and a similar problem would occur.

It was becoming very clear to us and our potential customers that there was still much to understand about the behavior of the Omega signals. We needed better insight into what was happening to the Omega signals at different times of the year as well as in different regions of the world. That need led us to seek out government agencies and airlines throughout the world that were willing to fly the system. We needed the information about the system's performance almost as much as we needed their business. It was a risk, exposing those problems to prospective customers, but we knew we would not have a market until we solved all the problems.

For years at various Omega symposia, presentations by operators, researchers, and equipment manufacturers would often refer to the latest signal anomalies they had uncovered. The equipment manufacturers would talk about how they solved a particularly nasty problem. The operators would talk about the next problem that the manufacturers would have to fix. The researchers would talk about the signal anomalies that still awaited manufacturers and operators alike. Sometimes the lane slips would cause errors of 100 miles. Those were the potentially dangerous errors and were given careful scrutiny when they occurred. Fortunately, that was not a frequent occurrence, but it happened often enough to keep engineers in several companies fully occupied.

When recounted in just a few paragraphs, those problems appear horrendous. In reality, they occurred over several years, but that made them more difficult to solve. However, their infrequent nature did allow the marketing effort to continue, and usually the equipment worked well.

Potential buyers were testing the competing systems, from Canadian Marconi and Northrop, on flights throughout North America and over the North Atlantic. Most of the time the systems were navigating to an accuracy within 2 nautical miles.

In many respects the design of the Canadian Marconi system was more advanced than the Northrop system because its development started later. The principal hardware difference was the computer. The Canadian Marconi Woodhouse computer was a more up-to-date design and, we thought, was better able to cope with the complex calculations that an airborne Omega Navigation System has to make.

The Northrop system, however, had one significant advantage: it was one of the first airborne systems to use the new concept of Kalman filtering in its software program. It used the Kalman filter to estimate position through a complex series of analyses of all the received Omega signals. The Kalman filter combined the statistically more reliable signals so that they had a stronger influence than the weaker signals in the final calculation of position.

That method of calculating position gave the Northrop system some advantages over the Marconi system in the sparse Omega signal environment. The Northrop system was performing well in Navy tests, and so, in the final analysis, we were competing in a market where we did not have a significant technical advantage over our competitor's product.

Both companies were pioneering the development of airborne Omega Navigation Systems. Our success or failure would determine whether or not the world would have a new, low-cost, long-range navigation system. The competitive situation prevented the companies from formally exchanging information on new signal irregularities we were experiencing during flight tests. However, the engineers were informally exchanging information at the various Omega symposia.

The U.S.-based Institute of Navigation had hosted the first international symposium dedicated to the Omega System in November 1971, one year before I joined the company. The symposium, which was held in Washington, DC, had attracted 350 engineers, scientists, government officials, and aircraft operators, which was a large attendance for such a specialized symposium. The event helped to stimulate interest in the Omega System, but 12 months later the prospective users still considered the system to be in the research phase.

The symposium also helped to unite the Omega community, which included potential users of the system, scientists, engineers, marketing professionals, and government officials. Over the years, it would expand, and despite fierce competition in the marketplace, we became a closely knit group of people. Our relationships were rarely antagonistic. At the personal level, we respected each other's professionalism; we also understood that we were being loyal to the firm or agency for which we worked. We tried to put competitive considerations aside as we worked together to improve the Omega System for all users; individual rewards would come only if we pooled our resources to sort out collectively the problems that we all were experiencing.

In 1971 the Northrop AN/ARN-99 airborne Omega Navigation System had undergone a formal technical evaluation at the Naval Air Test Center near the Patuxent River in Maryland. The lead investigator for the evaluation was Charlie Sakran, a bright young engineer who is still one of the Navy's principal authorities on naval aircraft navigation systems. As we worked with Charlie, we came to admire him for the depth of his technical knowledge and objectivity. His talents have earned him a prominent position in the international Omega community.

The U.S. Navy intended to install the AN/ARN-99 on its P-3 maritime patrol aircraft. However, by 1972, it had not started the formal procurement process. Canadian Marconi still had a chance to catch up with the lead that the Northrop Corporation had established.

My first two months at Canadian Marconi, in the early autumn of 1972, were a period of intense study. I had to assimilate as much as possible about our product, its potential market, and the Omega System. Within two months I had visited Wright–Patterson Air Force Base in Ohio, the U.S. Naval Air Systems Command in Washington, DC, and the Canadian National Defense Headquarters in Ottawa.

The North American Engineering Company helped Canadian Marconi Company with its U.S. military marketing. Warren Scott had set up that company for the sole purpose of representing foreign companies, mostly Canadian, within the U.S. military establishments. Headquartered in Washington, DC, the company employed full and part-time representatives in several places in the United States.

Most of the time I would travel on my own, even in those early months on the Marconi Omega team; we were and remained a lean operation. The North American Engineering Company representatives were particularly useful because they could introduce me to the people I had to know. They also had the market background knowledge that I lacked.

Canadian Marconi also had part-time representatives throughout the world. Most of those people worked for a commission on sales. In all of my years marketing throughout the world I made maximum use of our local representatives. We also tried to keep them informed every step of the way.

A common mistake of many marketing organizations is that they do not communicate frequently with their onsite representatives, including their trade consuls, the government employees located at consulates and embassies. That is an error in judgment. Those people, if kept well informed of the company's plans and goals, can provide valuable market information and make marketing trips more efficient. They also can advise on the most appropriate timing for a trip. Timing is important, not only to prevent wasting time and money on unproductive trips, but also to meet customers when they are most receptive.

Often companies make a double error in judgment. They do not keep their representatives informed; then they call upon them to help at the last minute when there are problems or the company is about to lose a competition.

By our first Christmas in Montreal, it was obvious that Don had been wrong when he had said to me at the Farnborough Air Show, "Well, you've got an easy job. The Omega System will sell itself." It was just as well that I was a neophyte marketing person. If I had known how much work lay ahead, I might have decided it would be easier to change careers and tried to be a brain surgeon.

4

The Search (for a Customer) Continues

Success is spelled W - O - R - K.

One company rarely has an absolute monopoly in a market. Competition keeps people and companies alert and helps them avoid the risks associated with complacency. It forces companies to hone their marketing skills and to look for more efficient ways of conducting their business. Competition also forces a seller to listen to the customer and to respond better to the customer's real, not presumed, requirements.

During the early years of launching our airborne Omega navigation product line, competition came in several forms. We were in a head-to-head competition with Northrop, but we both faced competition from those companies that were selling the more expensive but proven Inertial Navigation Systems. Later, five more companies entered the market with airborne Omega Navigation Systems.

Another unlikely competitor was evolving and eventually became formidable, particularly in the general aviation market. The company started in the garage of a California doctor, who was also an electronics hobbyist and a pilot of small aircraft. Dr. Frugenfeld and a small team of electronic wizards found a way to use the U.S. Navy's very low frequency communication signals for navigation. He generated enough interest in his concept, and he went ahead and designed a marketable navigation system. During the early 1970s Dr. Frugenfeld established Global Navigation Systems Inc. and set about selling his system to the general aviation and business aviation communities. He called his first product the GNS-100. It was cumbersome to operate, but it was usable almost anywhere in the world. The Omega community in the meantime was still waiting for the Omega stations.

The problem with Global's system was that it used military communication signals, unlike those of the Omega System, that were dedicated to navigation.

Although the Navy had no objection to others using the communications signals, it reserved the right to shut down a station or change frequencies for security reasons. The success of Global's product divided the very low frequency radio navigation world into two camps: the Omega purists and the proponents of navigation using the communication signals. Over the years we had many heated debates.

One such debate occurred at a conference in Ottawa in 1974. By then another U.S. company, the Communications Components Corporation, had a product to compete with Global's. That product was called the ONTRAC 1.

I was one of the last people to present a paper. Several papers preceding mine had dealt with the navigation technique that used the U.S. Navy communication signals. Their presenters had been outspoken, but not entirely factual, in their opposition to the Omega System.

I rose to speak: was I the only proponent of the Omega System in the room? For the thousandth time I argued that the U.S. government had designed the Omega System specifically for navigation purposes, whereas the other systems used signals that were intended only for military communications.

In my paper I wrote the following:

> It is important to note that the Omega Systems derive navigation data from fully legitimate signals intended and provided for navigation use. This is particularly important to airborne users where a dedicated certifiable system would seem to be essential. These signals are used in a determinate way which can be supported by analysis and the errors are bounded. The 10 kHz frequency band was set aside for Omega only after some extensive studies in the early '60s. This early work is still valid, although this aspect seems to be forgotten occasionally in some circles.

The preliminary rounds in the battle were now over. The protagonists were the Omega purists vs. the practicalists, who were promoting the use of the communication signals for navigation. Those of us in the Omega corner had two primary concerns. Obviously, the prospect of increased competition concerned us; the GNS and ONTRAC systems were more expensive than the Omega System, but they had signals to use. The Omega ground station network was still not complete.

Our biggest concern, though, was that because the U.S. Navy had complete control over the communication signals, the industry could not rely on their availability all the time. The air transport regulatory agencies in Canada, the United States, and elsewhere shared our concern. While they could not prevent aircraft operators installing the GNS and ONTRAC systems in their aircraft, they did not allow the systems to be the primary means of navigation. Regardless of those limitations, many of the early users trusted those systems because they worked effectively most of the time.

Over the next few years the regulatory agencies, industry, and the U.S. Navy held several meetings to try to resolve the dilemma. Understandably, the U.S. Navy maintained the position that it reserved the right to manipulate the signals as necessary. Eventually, the Navy agreed to provide advance notice if it intended to change any of the carrier signals, and so the regulatory agencies formally approved the carrier signals' use for backup navigation. The Navy very rarely changed the signals or temporarily shut down a station because it had ships, submarines, and aircraft that relied on those signals, and so in the end, both sides won. General and business aircraft operators bought the GNS and ONTRAC systems, and initially those sales far outnumbered sales of the Omega Navigation System in that segment of the aviation market.

Finally, in 1976, the Omega purists acknowledged the benefit of using the communication signals as an adjunct to the Omega signals, and so we added that capability to our systems. That feature proved to be invaluable to ensure early sales of our Omega Systems.

While we kept a wary eye on Global Navigation Incorporated, we had other more pressing problems. It was becoming obvious that the U.S. Air Force would not be buying Omega Systems for some time. It was also obvious that we were going to have a very difficult task convincing the U.S. Navy to abandon Northrop. We needed to look elsewhere for our launch customer.

Early in October 1972, Val Roseberry, our group manager, and I visited staff responsible for navigation systems at the Canadian National Defense Headquarters in Ottawa. We tried once more to convince them to fit Omega Systems in their aging Argus aircraft, the maritime patrol airplanes that they planned, eventually, to replace. The Canadian Armed Forces, realizing now that a replacement aircraft was still several years away, were willing to reconsider our suggestion. However, like everyone else, they wanted to test the system first.

Defense Headquarters tasked their Maritime Proving and Evaluation Unit to conduct the flight test. This unit operated from the Canadian forces base in Summerside, Prince Edward Island. During the next two months, Ormee Chamberlain and Frank Gagnon, two of our first Omega field engineers, traveled to Summerside to help the Canadian forces personnel. First they had to find a suitable location for the antenna on the Argus aircraft. Then they installed the cable harnesses together with the equipment racks and the three units that made up the CMA-719 system. By Nov. 30, 1972, they had checked that the system was ready for the flight tests. They also had trained the crews to operate the system and had helped write a flight test program.

The trials took place between November 1972 and March 1973. After the flights in the Argus, the Canadian forces transferred the system to a C-130 transport aircraft. We had the attention of the Canadian Armed Forces.

In March of 1974, at the invitation of the Canadian government, I presented a paper to a special NATO panel meeting at NATO headquarters in Brussels. Its title was "The Use of Omega in the Airborne Role Today—Canadian Armed Forces ARN-115 Trials." Long complicated titles were meant to convey a sense of authority on the subject.

The introduction to the paper gave some idea of the problems we were having convincing the market to use the Omega System. At the time of the Argus trials in 1972, only four experimental Omega stations were on the air. By the time of my paper in 1974, four full-power stations were on the air, plus one of the original low-power experimental stations. The system was in better shape from a signal coverage point of view, but it was not complete.

Nonetheless, I wrote, "Omega is no longer a science, it is a fact, and it is usable today. . . . The time has come for us to talk about the practical aspects and proven applications of Omega."

Bold words, but we had to use every opportunity to try to convince the airborne community that the Omega System was usable. At least it was usable in a little more than half the world, which was not much good for those operators who had worldwide routes.

One sentence in the paper pointed to another market problem that we had to overcome: "It is noteworthy that there will be no user charges for Omega."

The commercial airlines had to pay to use several of the ground-based navigation aids. One of those aids was a system called LORAN-A, which some airlines were using on transatlantic routes. The airlines resented having to pay for a government-operated system that the military also used. In the early 1970s they were stating their objections forcefully. They were on safe ground because they now had the option to use the new self-contained Inertial Navigation System. The airlines were afraid they would have to pay to use the Omega System.

Those of us who had a vested interest in the system encouraged the United States government to clarify its position. We had received several verbal promises that it would not levy any user fees, but we did not have that promise in writing. The airlines remained wary. Those of us marketing Omega receivers took every opportunity to repeat what the United States government was saying. To this day no user has been asked to pay for the service, and the Omega System remains one of the navigation bargains of all time.

Another perception that we had to constantly dispel was that many potential operators thought that the United States, or one of the partner countries, would shut down a station if there was a war. We had to point out that the system was not accurate enough to be a tactical system. It was a long-range general purpose navigation system providing navigation to within 4 nautical miles, which is not exactly the precision required to position bombs. In 1971 the management of the system changed from the U.S. Navy to the U.S. Coast Guard. That helped our cause because the Coast Guard comes under the civilian

Department of Transportation. It was not until early in 1975 that the U.S. Federal Aviation Administration (FAA) and other civilian regulatory agencies began producing regulations governing the use of Omega Navigation Systems. Then, finally, the civilian community acknowledged it as a bona fide system.

The flight tests of the Canadian Marconi Omega System by the Canadian Armed Forces were a success. During the four-month test, the Omega equipment was flown from Prince Edward Island to Greenland, Iceland, Norway, and Azores, as well as across Canada to Vancouver Island. Its navigation accuracy was between 1.6 and 2.8 nautical miles 95 percent of the time.

After the successful Argus flight tests, the Canadian government registered the CMA-719 with NATO. That allowed us to refer to the system by the all-important NATO number it assigned, the AN/ARN-115. That nomenclature gave the system credibility in military circles.

As I became more comfortable in my knowledge of the system, I traveled more often, alone and with colleagues, trying to secure a customer, any customer. Val Roseberry and I disagreed on what it took to market a product. Val gave me the impression that he thought I could sell using 10-cent stamps. Those were the days of inexpensive mail. I maintained that face-to-face, not letter-to-letter, contact was essential for success; it was and is.

Trips with Kieth Glegg, the vice president, were always hard work but lots of fun. Kieth mostly went to U.S. defense establishments. Together we covered the Pentagon, Wright–Patterson Air Force Base, Naval Air Systems Command, Tactical Air Command, and the Air Force's long-range transport Military Airlift Command. We visited all of the military aircraft manufacturers and the U.S. Navy. Kieth would always give the presentation in his energetic, easy, infectious style. I never tired of watching Kieth flip over presentation charts while talking without interruption.

For visual aids Kieth preferred to use large paper charts on an easel. They would appear unprofessional compared with today's computer graphic and photographic capabilities. Over the years I experimented with flip charts, transparencies, more commonly called viewgraphs, and 35-mm slides. I quickly gave up using charts for transparencies and 35-mm slides.

Transparencies are easier to produce, particularly if the presentation includes many charts with text. They can be shown in a room with the lights on. Being able to keep the lights on during a presentation is useful because it allows the presenter to maintain better eye contact with the audience. An illuminated room also makes it easier for people to ask questions. People are also less likely to fall asleep when the lights are on, but then the presentation should keep them awake.

I struggled for years swapping between the two media. Finally, I realized that transparencies are better for working on marketing presentations when the audience contains less than 50 people and I want to encourage a two-way discussion. On the other hand, 35-mm slides are better for more formal presentations, such as at a conference with 200 or more people in the audience. This is particularly true when I want my presentation to be smooth and I do not want to fumble as I place vugraphs on the projector or to say "next chart please" every 2 minutes.

On many of our marketing trips we would have to handle over 100 pounds of CMA-719 demonstration equipment. Several times Kieth Glegg and I staggered through the seemingly endless corridors of the Pentagon on our way to the office of a multistar general. Once in the office, Kieth would resume his role of vice president, while I scrambled around preparing the demonstration equipment.

Other than a brief flirtation with the airline market in the 1950s, Canadian Marconi had directed all its effort in the military market. In that it had been very successful, selling both ground-based communications systems and aircraft Doppler Navigation Systems.

I was more comfortable in the commercial rather than the military market, and so it was natural that I looked to the airlines and commercial airframe manufacturers for a possible launch customer for the CMA-719. At that time the airlines were buying, or preparing to buy, new aircraft. Those included the Boeing B-727, B-737, and B-747, as well as the new Douglas DC-10. Lockheed was in the process of designing the L-1011.

I made several trips to the West Coast. Our marketing presentations still included promoting the company, explaining the Omega concept, because it was still unknown to many engineers, and describing our product.

Frequently, we had to answer questions such as these: "How serious is Canadian Marconi about entering the commercial airline market?" "What do you know about the airline market?" "Are you prepared to provide us with the kind of support we require?" "Do you know what kind of support the airlines require?" The questions came in rapid succession. In truth, we were not prepared: we thought we knew what the airlines wanted. I had some idea after my three years at Boeing, but we had not even started to understand what it would take to break into that market, succeed, and stay in it.

We had the support of the company management to pursue the commercial market; it would actually be more accurate to say that management did not discourage us. The company's primary focus was still the military. As a result, our management did not view the commercial market as the place where the company would invest a lot of resources. That would change, in time.

We Still Needed a Launch Customer

We still needed a launch customer for the CMA-719. Despite the amount of time that we spent visiting prospective commercial customers, we spent more time with military organizations. The U.S. Navy, the U.S. Air Force, and the Canadian Armed Forces were still at the top of our list.

When, four years earlier, the U.S. Navy awarded a contract to the Northrop Corporation, it was only for the development of an airborne Omega Navigation

Teaming a Product and a Global Market

System and a few sets for test purposes. The Navy intended to hold a competition later for a production contract.

By 1971 Charlie Sakran, and his people at the Naval Air Test Center, had tested the Northrop system in a Lockheed maritime patrol aircraft, the P-3, as well as in helicopters and submarines. The Navy loaned systems to the U.S. Air Force, which flew the system in some of its jet transport aircraft. In addition, Northrop had persuaded Continental Airlines and World Airways to test the system in Boeing B-707 and Douglas DC-8 aircraft.

Undaunted by Northrop's lead, we continued to visit the U.S. Navy's Naval Air Systems Command. We tried repeatedly to convince it that our CMA-719 system was better than Northrop's.

It is a matter of some debate whether or not our system was better than the Northrop system. Twenty years later I can be more objective than I was at age twenty-seven, when I was desperate to secure our first major sale. The Northrop system with its advanced Kalman filter software was, in many respects, better than the Marconi product.

Those early visits to the Naval Air Systems Command were the beginning of a long and close working relationship that I developed with Irving Lublin, the Navy's airborne Omega project officer. Irving held the door to our entry into the Navy. Befriending him, therefore, became a necessity but not a chore. Many years later I gave a speech during an evening honoring Irving and his wife, Betty, for his work in the International Omega Association. It was a speech that I was happy to give. During it, I publicly admitted to Irving that at first I had befriended him hoping to gain an advantage to win the Navy Omega contract. I told the audience that during all the years I worked with Irving he only once came close to giving us that advantage. During the final stage of the bidding process, he said, "Graham, whatever you do, make sure you keep your pencils very sharp." That was it; "keep your pencils sharp." He was telling us that we had to have the lowest price to win. No special inside information, no information on what the winning price would be, nothing except use a pencil sharpener.

Whenever I give marketing lectures, I tell people that you must really want to win to succeed with a bid. However, you should not want to win at any cost. The bid should be under profitable conditions, and the contract won honestly. No contract is worth a term in jail.

Our persistence with the Navy finally paid off when it agreed to test the CMA-719 alongside the Northrop system in a P-3 maritime patrol aircraft. Once again the Canadian Marconi field engineers built cables for the installation. Then they went out to help Navy technicians install and check the system in the aircraft. The flight tests took place early in 1973. Our field engineers drew lots to select the lucky person who would miss the Montreal winter. The principal locations of the test flights were the warm climates of Texas, Florida, and

Bermuda. The official explanation was that the sites were chosen on the basis of available Omega signal coverage and aircraft tracking ranges, not the weather.

On average, the two systems performed the same. However the Northrop AN/ARN-99 did perform better than our system on a small percentage of occasions, especially when the signal coverage was marginal. The U.S. Navy concluded that the slightly better performance of the Northrop system was due to the sophisticated Kalman filter in its design. The performance difference was not much, only one-third of a nautical mile within a range of 2 nautical miles, most of the time. However, the situation was highly competitive, and we were trying to disrupt a program that was already under way. It was not surprising then that the Marconi system would have had to perform significantly better than the Northrop system for the Navy to be prepared to change.

For several months afterward we made repeated visits to Irving Lublin and others in an attempt to convince them that the Marconi system was better in several respects. It was less expensive than the Northrop system. We had calculated that its support and maintenance costs would also be lower. However, all our attempts to break into the Navy were of no avail.

What we had not comprehended fully was the depth of the working relationship on that program between Northrop's and the Navy's engineers. They had learned to work together. We also had not acknowledged the commitment that the Navy had made to the Northrop system. If that was not enough, the Northrop system also worked just fine, and there was nothing we could do about it. When it came time for the Navy to take its Omega Navigation System program from development to production, it awarded Northrop the contract.

We learned from our first Navy experience. The flight tests gave us some valuable information about the performance of our Omega System. We also learned about relationships between suppliers and customers. In the commercial market, that is called product loyalty. In the military market, it is more a question of the government agency not wanting or even being able to change suppliers once a program has started. To do so would mean having to recompete the entire program. That could introduce years of delays and void years of preparatory work.

At the end of 1972, we did not know the outcome of the Navy program. What we experienced was an increasing interest in the Omega System. That maintained our optimism, and we welcomed any news that might give a boost to the Omega market.

One such piece of news came out of the FAA. It announced that in 1976 it would close down the LORAN-A navigation ground stations. A government agency usually makes such announcements several years in advance of the event. Often such changes require operators to buy new equipment, and procurement and installation, in a large fleet of aircraft, can take several years.

The World War II vintage LORAN-A ground stations were providing navigation signals over the North Atlantic, to South America, and across the

Northern Pacific. Most of the older long-range aircraft, such as the Boeing B-707 and the Douglas DC-8, still navigated with LORAN-A and Doppler equipment, whereas the newer aircraft, such as Boeing's B-747 jumbo jet, used Inertial Navigation Systems.

The planned shutdown of the LORAN-A chain presented the operators of the older aircraft with a problem and those of us in the Omega community with an opportunity. The airlines had to decide how much longer they would keep those middle-aged aircraft. Then they had to decide if it was worthwhile equipping them with expensive Inertial Navigation Systems. If we could get the Omega System working in time, it would be a viable, and much less expensive alternative.

The planned shutdown of the LORAN-A stations within four years was a major contributor to the heightened interest in the Omega System. We built a small preproduction run of 15 CMA-719 systems. We had sold four to the Canadian Ministry of Transport for installation on the Nordair-operated Electra aircraft, which monitored ice conditions in the Arctic. Our engineers had three systems in the Omega Lab to support their design work. We had loaned one system to Marconi Avionics in England for marketing demonstration purposes. We quickly assigned the other seven to potential customers for flight tests in their own aircraft.

In the avionics business, most customers, regardless of their size, want to test fly a system before they buy, especially a new or expensive system or a system that performs a critical task. Often we would find ourselves involved in a flyoff against our competitor's equipment. Such is the nature of this business. A company not willing to commit the considerable financial and human resources to that type of marketing activity will not succeed. The same is true for many other specialized businesses with high-technology products.

Our visits to the U.S. Air Force, the Royal Air Force, Boeing, Air Canada, the Ministry of Defense in France, and others paid off in ways that we did not quite expect. They all wanted to test fly a system—at no cost.

In May of 1973 we were struggling to support eight separate flight evaluations of the CMA-719. Sterling Airways in England was flying the system in a Caravelle. The Canadian forces moved its system from the Argus to a Lockheed C-130 transport aircraft. The U.S. Navy was still testing the system alongside the Northrop system in a P-3 somewhere near Bermuda. The Royal Air Force once again had a system in its Comet test airplane. The U.S. Air Force had a system installed in a Lockheed C-141 jet transport flying over the North Atlantic. Boeing installed a system in its demonstration long-range maritime patrol aircraft, the type they hoped to sell to the Canadian Armed Forces.

In addition, we had convinced the FAA to test the system in its KC-135 aircraft. We did not expect the FAA to buy any systems. We simply wanted to prove that the Omega System was viable and that the agency could safely sanction the system's use for navigation within its area of jurisdiction. Those flight evaluations were an example of the type of investment that we and others made to help launch the Omega market. The success of the FAA tests gave it enough confidence to continue considering an Omega Navigation System as an alternative to the LORAN-A equipment being used. Except for the FAA, we had high hopes that the other evaluations would produce some near-term business. On the basis of that interest, Kieth Glegg finally approved the manufacture of 100 CMA-719 systems. That represented a considerable investment and risk. However, we concluded that we would need to be ready with systems off-the-shelf to secure the early orders.

Soon afterward we lost, or rather did not win, the U.S. Navy Omega production contract, which was awarded to Northrop. Kieth Glegg, who reveled in his command of the English language, would correct us whenever we talked about having lost this or that contract. Fortunately, that was not too often. "How," he would say, "is it possible to lose something that you never had in the first place? You perhaps did not win, but you most certainly did not lose."

The planned shutdown of the LORAN-A chain gave the U.S. Air Force the same problem as the airlines. It had many C-130 and other transport aircraft that would need a new navigation system. That gave us confidence that it would buy Omega Systems. It had even started flight testing our system.

Canadian Marconi had been particularly successful with the U.S. Air Force in the past. The company had created the Avionics Division as a result of its successful pioneering Doppler development work and substantial sales to the Air Force. We knew the Air Force, and we used our many contacts to launch that Omega flight test program. We considered it unlikely that the Air Force would buy the Northrop system that the Navy had selected. In the past, in such matters, the two services had shown a preference to exert a certain amount of independence. That gave us a significant competitive advantage, or so we thought. Meanwhile, the large U.S. Bendix Corporation, a small U.S. company called Dynell Electronics, and others were beginning to take an interest in designing Omega equipment.

Two to three years is a short time for a military agency or airline to conduct a competitive procurement program and then outfit its aircraft. Airlines and military operators cannot afford to have their aircraft grounded for a long time while they install a new system. When a fleet retrofit of a new system is to take place, they wait until they have to ground an aircraft for a scheduled overhaul. An airline usually only needs to overhaul an aircraft once each year. As a result, it will take 18 months to retrofit a new navigation system into a fleet of 100 airplanes.

The military procurement cycle is long and complicated, particularly in the United States. In its basic form, it starts with a user command, such as the U.S. Air Force Military Airlift command, announcing a requirement for a particular system. In this case, a long-range navigation system was slated to replace the soon-to-be-obsolete LORAN-A equipment. The command documents that requirement in a report that makes its way through various layers of bureaucracy. In this case, the requirement first went to Wright–Patterson Air Force Base, where engineers determined what type of system would satisfy the requirement. The requirement and the proposed solution then make their way through more layers of bureaucracy, and eventually a cost estimate is added. The Pentagon then analyzes the entire package. If it is convinced that the requirement is justified, it approves and includes it in the Air Force budget submission to the president and subsequently to Congress. Once Congress has approved the budget, the Air Force can go ahead with a competitive procurement. The competitive procurement may first involve a product development program and later a competitive test flight program. The process can easily take three or more years.

The smart company begins at the beginning of the process and helps whenever and wherever possible. Often the military agency will ask companies showing interest in the procurement to review specifications and other documents to ensure that what it is asking for is realistic. That type of marketing requires a company to invest much time and effort. However, that way a company can keep in touch with the requirements and schedule as they evolve. In the end, you win, or do not win, based on your final bid and how closely your product and its cost meet the customer's specification.

The U.S. Air Force flew the Canadian Marconi CMA-719, the AN/ARN-115, as it was now called, Omega System for more than 300 hours. Throughout the flight test program, the system navigated to within 3 nautical miles, which was within the expected accuracy of the system and well within the requirements for navigating safely across the Atlantic Ocean. The accuracy was, in fact, far superior to most other long-range navigation systems.

To capitalize on the success of the Air Force test program, we began to travel more often to the Military Airlift Command. We wanted to make sure that we fully understood its requirements. We went to Wright–Patterson Air Force Base to encourage its engineers to recommend that the Air Force use Omega Navigation Systems, not Inertial Navigation Systems. We kept the Air Force Logistics Command apprised of the likely procurement program. It would become actively involved later, once the technical evaluations were completed. We visited the Pentagon and the Lockheed Aircraft Company, the manufacturer of the two primary Air Force transport aircraft, the C-130 and C-141. In short, we visited and revisited anyone who would have some involvement in the procurement when it occurred. To succeed, we knew we had to try to prepare ourselves for every eventuality and that we had to keep on top of the program.

The Omega station situation plagued us for several years. The continuously slipping schedule eroded customers' confidence in the system and represented our single largest hurdle in launching the system.

We were conducting our marketing in a rather fragmented fashion. We approached both military and airline operators trying to secure a customer, any customer. We lacked a focus and a definitive strategy. We were feeling our way, trying to figure out our best route as we went along.

We eventually developed a coherent strategy, but only after our primary market became more obvious. More important, we had the common sense to realize that we had to understand the implications of our strategy. We knew we had to assess the financial and personnel investment that the strategy represented. We also had to match the strategy with the company's goals and priorities and to understand what infrastructure we would need. It was only after all this came together that we became successful.

In the meantime we had to learn the hard way. Although the company had been very successful with its Doppler Navigation Product line, the Omega team was dealing with the launch of a new product. We were also involved in the launch of an entirely new concept in navigation. Some of the lessons learned with the Doppler product line applied, but many did not. We were learning by trial and error.

In October 1973 the Canadian Armed Forces Maritime Proving and Evaluation Unit published its final report on the results of the CMA-719 flight tests. The results were favorable. They proved, conclusively, that the system would meet the Canadian forces maritime patrol navigation requirements. They also confirmed that the Omega System would be a suitable replacement for the LORAN-A system.

I had been visiting our National Defense Headquarters for some time trying to conclude a sale to the Canadian forces. A young captain, Earl Price, was in charge of the project. He knew how to deal with the government bureaucracy and had a reputation for getting the job done. Earl was one of the few officers to be promoted, first to major then to lieutenant colonel, as soon as he became eligible. I have a lot of respect for Earl, and I am grateful to consider him a friend.

Earl appreciated the benefits of the Omega System from the beginning. The Canadian forces needed a new navigation system for several aircraft. There was pressure from Litton Canada, a subsidiary of U.S.-based Litton Industries, which was trying to convince the Canadian forces to buy its Inertial Navigation Systems. Earl led the effort to convince the hierarchy in the Canadian forces to buy Omega Systems, a campaign that he eventually won.

5

Where Do We Go from Here?

*The great thing in this world is not so much where we are,
but in what direction we are going.*
—Oliver Wendell Holmes

In August 1973, I went on my first sales trip to South America. I had set up that trip to coincide with an aerospace show taking place in São Paulo, Brazil. I went first to Chile, then to Argentina, and finally to Brazil.

I recall being somewhat apprehensive about the trip. I knew very little about South America, except what I had learned in school geography lessons. We had not even covered South American history in any detail. Also, I was concerned because I really did not know the business environment in South America. The company had marketing agencies in those countries, and I relied heavily on their staff. They arranged our meetings and guided me in our business dealings.

I did not expect to close any kind of business deal during the trip. It was what we called a fishing expedition. The primary purpose was to contact prospective customers and to learn about their requirements. Then we could determine if there was any real potential for business in the future. As in many businesses, the road to opening a market can be a long one. Successful business people understand the need to nurture a market. An overanxious salesman can easily discourage a potential customer. Many salesmen can open a market but are not capable of closing a deal. The trick is knowing when to tread carefully and when to be more aggressive.

Usually high-technology products are sold to sophisticated customers. It is not like selling a pair of shoes off the shelf. No amount of selling will convince the customers to buy the product. They have to be convinced that the product and the company meet all of their requirements and that the proposal is competitive. The seller must always be attentive and responsive to the customers' queries, not just during the final negotiation stage.

Whenever I journeyed to a country that I had not visited before, I made a point of learning a little about the geography, politics, and customs of the country. It is good marketing to be able to converse with your potential customer on several subjects. Customers appreciate the salesman who shows a genuine interest in their country and who is sufficiently knowledgeable to ask some sensible questions. It is bad marketing, however, to voice an opinion about the politics of your host's country.

Mostly, I was concerned about going to Chile. A bloody military coup a year earlier had resulted in the ouster and alleged suicide of the Marxist president Salvador Allende. A right-wing military junta was now ruling the country.

Argentina was in the second Juan Perón era. During the 18 years since Perón had last been in power, several military and civilian governments had ruled Argentina. Juan Perón died in 1974. His wife Isabel succeeded him until she was overthrown by another military junta in 1976.

A military government had ruled in Brazil since 1964. During the 1970s the military government began a concerted effort to democratize the political structure of the country.

Military governments ruled in two of the three countries that I was to visit. The third had a civilian government that had been in power for only one year. Naturally, I was a little apprehensive about the trip.

My first stop was Santiago. Tom Bertie, our local representative was at the airport to greet me. Tom was the personal assistant of the manager of the company that represented Canadian Marconi, among others, in Chile. He was in his mid-twenties, amiable, and anxious to please. During the next three days we held several meetings with Lan Chile, the country's national airline, and various departments in the Chilean Air Force.

At each meeting I gave the Omega presentation using 35-mm slides, not the paper charts favored by Kieth Glegg. I had improved the presentation over the months. It started with a brief outline of the Canadian Marconi company. A section on Omega theory followed, along with a technical description of our Omega Navigation System, the CMA-719. To conclude the presentation and to show that the system really did work, I had slides summarizing the results of the various test flights. Because we still did not have a significant contract, we could only talk about the equipment's performance, not its customers.

For several years we included Omega theory in most of the presentations. It was essential for our potential customers to understand how the system worked so that they could appreciate its many advantages, as well as its limitations. Those theory sessions also allowed us to correct any misconceptions that our potential customers had about the system. A lot of misinformation was going around. We were learning the difference between marketing a new product in a well-known line of similar products and marketing a new product concept.

The Chilean Air Force showed interest in the system, as did Lan Chile. The airline knew that sooner or later it would have to replace the LORAN-A systems in its B-707 aircraft. We still had the problem of convincing the Chileans, like most other people in the aviation industry, that the Omega System was practical. The delay in the building of the transmitters was only one of our problems. They had heard of the occasional, and unexplained, Omega signal anomalies that were causing large navigation errors on some routes. They thought that the U.S. military would shut down the stations if a war broke out. Also, the Chileans were aware that the FAA, which regulates air transportation safety and navigation in the United States, had not yet approved the Omega System for navigation purposes. In short they wanted someone else to buy airborne Omega Systems before they did.

Despite those expressions of uncertainty, I felt that the visit had been successful. I was sure that a market for the product would eventually develop in Chile. It was up to us to nurture that market. We did. A few years later, the Chilean Air Force, Lan Chile, and other government agencies in Chile bought many of our Omega Navigation Systems. However, they did not buy our first-generation system.

The Chileans had been friendly and positive. At least they had left me with a sense of optimism. The Argentineans on the other hand, were more serious, and it was difficult for me to judge the market potential in Argentina. Aerolineas Argentinas, the national airline, and the Argentinean Air Force appeared to have a requirement for our type of product, but they did not express an opinion. Argentina turned out to be a very difficult market for us to penetrate. Despite many follow-on visits and flight tests of our equipment by both Aerolineas Argentinas and the Argentinean Air Force, we had little success.

I was fascinated by my first visits to Santiago and Buenos Aires. The foothills of the Andes began just a few steps from my Santiago hotel window. The Andes dominate that country, which is about 2500 miles from north to south and only 250 miles at its widest point. The Andes run the entire length of the country.

I sensed that Buenos Aires was more Latin than Santiago. In a pedestrian thoroughfare, a short distance from the hotel, groups of people earnestly debated, I assumed about politics, in loud voices. A crowd stood outside a newspaper printing house. The company posted the latest news on large bulletin boards in its window. People eagerly gathered around the window to read the latest news. Elsewhere old men sat playing chess. Bars and bistros spilled Argentinean music out into the night air.

My next and last stop on my first marketing trip to South America was Rio de Janeiro, Brazil. It was love at first sight.

People give countries their character. A visitor, particularly a vacationer, may be in awe of ancient ruins and man-made wonders, such as the pyramids of Egypt. However, for the business person, it is the people who really count.

A smiling Brazilian named Antonio Miranda met me at the Aeroporto Internacional do Galeão, Rio de Janeiro's international airport. Antonio worked for the company that represented Canadian Marconi and other Canadian companies in Brazil. He had been an ace pilot in the Brazilian Air Force and for several years had been a member of the Brazilian Air Force aerobatics team, which gave him a certain amount of notoriety. After leaving the Air Force, Antonio had been a talk show host on Brazilian television. He was an aviator at heart and so returned to the industry that he loved. His job allowed, in fact, it demanded, that he spend most of his time meeting with Brazilian military, airline, and airframe manufacturer personnel, an environment in which he was and continues to be most comfortable. His attachment to the industry shows and makes him a most effective marketing representative.

Above all else, Antonio is a professional. He does not ignore a customer after a sale. He understands that satisfied customers represent repeat orders. As important, he ensures that the customers are satisfied with the product and the service they receive from the company that made it. Antonio and I became, and have remained, good friends, despite the 5000 miles between our homes.

The Portuguese colonized Brazil. The first and most obvious difference from its neighbors is that the 120 million Brazilians speak Portuguese, not a variety of Spanish. The Brazilians are also more racially integrated within the three ethnic groups that call Brazil home. The large majority of Brazilians are the native Indians and the descendants of the West African slaves and the Portuguese settlers. Also, a large population of German immigrants have settled mostly in the southern part of Brazil.

Brazil is a huge country, much larger than any of its neighbors, occupying 50 percent of the South American continent. The country is very rich in natural resources, and the 1970s was a period of massive industrial growth. This well-orchestrated growth made Brazil the leading industrial power in Latin America. I always found the people to be friendly and very professional in their business dealings.

Antonio, through his connections, was able to speed my progress through customs and immigration, even though I carried the usual 35-mm projector and other marketing paraphernalia. Then we drove to the hotel he had booked for me, which was the Oure Verde, a small hotel catering primarily to business people. After my first visit, I no longer had to check in with my passport and other official documents. Most times I did not even have to sign in. They said, "Welcome back to Brazil, Mr. Gibbs. Here is your room key. Let me help you with your bags." It was not that I was a special guest. They treated all return guests that way. It was just the type of treatment that a business traveler appreciates after a tiring 16 hours of travel from Montreal to Rio de Janeiro.

The hotel was located near the famous beach of Copacabana. Crossing the busy Avenue Atlantica from the hotel to the beach was a challenge, but once I was safely across, there were miles of sandy beach to enjoy. Residents of Rio de Janeiro, called cariocas, knew about power walking long before it became fashionable in North America. Every morning, it seemed as if all the five million inhabitants of Rio de Janeiro were on the promenade or the beach, engaged in a variety of physical activities, which made a walk on the long, wide beach more like a trip on an obstacle course. Most of the space was taken up with staked out soccer pitches, where hundreds of cariocas practiced the national sport before going to work. At night people crowded the many restaurants and cafes along the oceanfront. South of Copacabana, around a headland, was Ipanema, where the spectacle is repeated.

No visit to Rio de Janeiro is complete without a drive up the mountain of Corcovado. At the summit stands the inspiring statue of Christ with his arms outstretched. The statue dominates the skyline of Rio de Janeiro and is a reminder that this is a country of Catholics. Christ with his arms outstretched in a welcoming gesture is also symbolic of this country.

Cariocas seem to enjoy life and to find ways to cope with difficult situations. It is hard to imagine those people getting ulcers or allowing a situation to cause them stress. Antonio once told me a story that illustrates how the cariocas deal with a life that is not always easy for them.

During one particular winter, a very large pothole appeared in a downtown residential area. The hole was so big that several cars were damaged when they were accidentally driven over the hole. No amount of complaining and letter writing to the city maintenance department got the hole repaired. The hole remained unrepaired for 12 months. As the anniversary approached, the local residents conceived a plan to bring attention to their plight. They organized a lavish street party, with music and dancing, to celebrate the birthday of the hole. They invited the newspapers, and the radio and television stations to cover the event. The city was so ashamed of the publicity that maintenance crews filled the hole the next day. It was a practical solution to an aggravating problem, and the people also had a party.

Sadly, for many people who live in Rio de Janeiro life is not one long beach party. In the hills overlooking the city are shanty villages called favelas, where the poor of Rio de Janeiro live in makeshift huts or under canvas awnings. Many trek down the hills each day and become the city's street urchins and peddlers. Along the oceanfront and in the business districts children try to clean every pair of shoes in sight, in return for a few cruzeiros. Others squat on the sidewalks and peddle cheap souvenirs, trinkets, and puppets. Still others just beg, and a few pick pockets.

The first meetings Antonio and I had were at Varig Brazilian Airlines, the national airline. Many engineers and some pilots attended my Omega presentation. Afterward we held a meeting with Assis Arantes, the assistant to the director of engineering. Assis was another friendly and professional Brazilian. From our first meeting, I admired Assis for his professionalism, and his helpful and friendly manner. We liked and respected one another.

Varig had been keeping up to date with Omega events in the United States. Its managers, like those at Lan Chile, knew that one day they would have to replace the LORAN-A equipment on their B-707 airplanes. They were reluctant to buy the expensive Inertial Navigation Systems for aircraft that were already well past midlife. Assis confirmed that they were interested in the Omega System. However, like everyone else, they had doubts about its readiness for use by an airline.

Assis made it clear that Varig's engineers would want to test a system in an aircraft before they would even consider procurement. That was becoming a routine request. I readily agreed to an evaluation, although we were beginning to get so many requests for evaluations that scheduling equipment was becoming a problem. The cost of supporting so many evaluations was also becoming a problem.

In the years that followed, I visited Varig on many occasions, and our two companies enjoyed an excellent business relationship. Many Varig airplanes would come to rely on the Canadian Marconi Omega System. Technical problems cropped up from time to time, and one in particular proved to be our greatest challenge. However, the two companies always honored their commitments to each other. Canadian Marconi for its part remained solidly behind the product. When we encountered technical problems, our engineers investigated and solved them regardless of the cost.

Seventeen years later, in October 1990, Antonio Miranda wrote a letter to me:

Dear Graham,
Hope that you, Chris, and the kids are going well. The seeds that you planted in Varig are still sprouting. Next week CMC [Canadian Marconi Company] should receive the official "green light" to go ahead with a program of supplying around 50 each Alpha-Omega units [one of our later systems] for the Varig domestic fleet of B-727 and B-737.
Sincerely, I have no doubts that your participation in the past was very fundamental in obtaining all that present success.
Warmest regards!
Antonio

Canadian Marconi had already sold equipment to Varig for most of its long-range aircraft. Now the company was going to start selling equipment for its domestic fleet.

While I was in Rio de Janeiro, Antonio and I held meetings with the Brazilian Air Force. We flew south to São Paulo to attend Brazil's first aviation trade show. Brazil's remarkably successful airframe manufacturer Embraer is located near São Paulo, and so we used the opportunity to meet with several of its engineers.

The Brazilian government created Embraer as part of its industrialization program. That program made impressive progress in the 1970s. Embraer began by building under license foreign-designed aircraft for Brazilian needs. Very soon it was designing and building small private aircraft, then military trainers and medium-sized transport aircraft. It concentrated on turbopropeller aircraft and today enjoys a worldwide reputation and market.

Two months before my first trip to South America, Christina and I had begun to search for our first home. Just before I left, we found a house near the city. What remained was to try to buy it at a price we could afford. I went to Brazil and left matters in Christina's hands. Upon my return, I discovered that we were the proud owners of a four bedroom Victorian row house that needed major renovations.

It was the autumn of 1973. Christina and I moved into our first home, and we immediately began demolishing the kitchen that had not been upgraded in 80 years. Scuba diving was starting to progress from a recreational activity to a serious commitment. John Rogers had convinced me to start training to become a volunteer scuba diving instructor at the YMCA, and that activity was proving to be very time consuming. I started juggling Marconi demands, with helping Christina renovate our home, while finding time for scuba diving. We worked hard but also played hard.

At work we hoped that the trials of the CMA-719 by the Canadian Armed Forces would result in a contract. We had made contacts throughout the world and several evaluations were under way.

In July, we had finally convinced Air Canada to flight test the CMA-719. Air Canada had been about to buy Inertial Navigation Systems to replace the LORAN-A equipment in its DC-8 airplanes. In the end the airline concluded that the Omega System was not yet viable, and so it went ahead with its original plans and bought Inertial Navigation Systems, at twice the price of Omega equipment.

Sterling Airways in England had flown our equipment, and during the spring of 1973 the Royal Air Force had once again flight tested the CMA-719. The U.S. Navy had flown a system in a P-3 maritime patrol aircraft. We also agreed to a flight test program with the FAA in the United States. Boeing installed a system in its B-707 maritime patrol test and demonstration aircraft. At Canadian Marconi, we felt that the tests would help us to promote our product for Boeing's commercial airplanes; even Boeing was unsuccessful in the military patrol aircraft market.

We were trying to identify all the possible customers, despite our limited marketing resources and the few systems available for evaluations. We realized that to have any hope of success we had to cover as many market opportunities as possible. We were, after all, launching a new Marconi product that had the additional burden of being a generically new product to the world. We did not know whether our breakthrough would be with a military agency or a commercial airline.

Most of the major airlines were having to face the inevitable shutdown of the LORAN-A navigation system service and the resulting replacement of the LORAN-A aircraft equipment. The airlines with the largest fleets equipped with the LORAN-A systems were Pan American World Airways, Trans World

Airlines, KLM Royal Dutch Airlines, Varig Brazilian Airlines, and Japan Airlines. There were many others, but those airlines were the most prestigious and the ones most likely to influence other airlines.

At that stage there was no specification for the installation of Omega Systems in aircraft. We would need such a standard before we could expect the airlines to buy Omega Systems. Because no such standard existed, Canadian Marconi engineers had designed the CMA-719 Omega System so that it would fit in the installation for Inertial Navigation Systems, which was standardized. The airlines, however, were still being very cautious when considering the Omega System.

The airlines were not alone in their planning for the shutdown of the LORAN-A chain. Many military operators would also be affected. Military transport aircraft operate on many of the same oceanic routes as commercial airlines, and so they must conform to the same international regulations as the airlines. The U.S. Navy had protected itself with the Northrop contract for the AN/ARN-99 Omega Navigation System. The Canadian Armed Forces looked almost certain to buy our CMA-719 Omega Navigation System, and very soon. A big question mark, though, was what the U.S. Air Force was going to do when the LORAN-A stations were shut down. It had a large fleet of transport aircraft that would need a new navigation system. More important, it had already expressed an interest in Omega.

A contract with one of the U.S. military agencies also virtually guaranteed sales to many other foreign governments. The earlier selection of a Canadian Marconi Doppler Navigation System for many U.S. military transport aircraft and helicopters had proven that fact. We were confident that we had an advantage with the U.S. Air Force because it had already successfully tested the CMA-719 in a C-141 transport aircraft. All we needed was the Air Force to launch a formal procurement activity.

Britain's Royal Air Force, the Belgian Air Force, and other military operators with C-130 aircraft were also on our list of potential customers.

Despite all of that, 1973 closed with us asking ourselves, where do we go from here?

6

A Customer at Last

Many of life's failures are people who did not realize how close they were to success when they gave up.
—Thomas Edison

It was January 1974, six years after Canadian Marconi had taken those first tentative steps into the Omega business. We now had a product but still no launch customer.

A month earlier, in December, the U.S. Coast Guard had completed the conversion of the Omega station in Norway from a low-power experimental station to a full-power station. Norway's station thus joined the Omega station in North Dakota as operational. The stations in Hawaii and Trinidad were transmitting, but as experimental stations. During 1973 the U.S. government had decided not to convert the station in Trinidad into a full-power station, deciding instead, with the agreement of the Liberian government, to locate a station near Monrovia. The change of location from Trinidad to Liberia re-emphasized that the Omega ground network was still subject to some changes—though work continued on the other stations in the chain, the question was when the job would be completed.

Although we still did not have any firm customers, 1973 had been a successful year. We had done our best to positively position ourselves for what we still expected to be a major product launch. Three airlines, four military agencies, the FAA and Boeing all had conducted flight trials with the CMA-719. Furthermore, the trials had proved that the system worked, at least, it worked in the areas we had flown—the North Atlantic, Europe, the eastern seaboard of the United States, and most of Canada.

Meanwhile, in the Omega Lab Peter Gasser and the engineers were still trying to solve several aggravating design and production problems. Each flight test gave us more information about the behavior of the Omega signals. Some data we had not predicated, and so the engineers had to make many changes to the system's software program. At times it seemed as if the 20-plus engineers working on the CMA-719 had a job for life. It was obvious the company could not maintain such a group of engineers on one product, particularly if that product had no sales.

Readers may be surprised that, even parenthetically, I have suggested that at the end of 1973 we were asking ourselves where to go from here. That was nearly six years after the start of the development of the product. Surely by now we could answer such a simple question? If not, then why were we all not fired? The reason was that we did not have a focused marketing plan. What we had was a rather wide-ranging one. We did not know and had no way of knowing what our primary markets would be.

By now we were convinced that many sales prospects existed in both the commercial and military sectors of the market. However, the market still needed concrete evidence that the Omega System was reliable and usable on a worldwide basis.

We were continuously improving our product as we gained a better understanding, through actual flight experience, of the real Omega signal environment. Another restriction was that our potential customers needed a real reason—a firm requirement—to fit their aircraft with Omega. We needed an event such as an announcement that the LORAN-A navigation chain definitely would be shut down. Up to that point the FAA had announced it was only considering such a move.

We all were working long hours. I was spending more than 50 percent of my time traveling within North America and overseas. Many of the senior engineers, such as Peter Gasser, were also traveling to meet potential customers. In such an environment, we could not have, nor should we have had, a focused marketing plan.

One of the strengths of the Canadian Marconi Company is that it does not have a large corporate bureaucracy. No group of people sits isolated in an ivory tower, preparing strategic plans. We had a corporate strategic plan for the company as a whole, but it was more global in nature. The company's corporate management delegates short-term strategic planning and the preparation of the even shorter-term tactical marketing plans. Short-term strategic planning is carried out at the division level. Tactical planning is carried out within the product manager's organization, by the people who work in the product's market every day. Management recognizes that those people are the most knowledgeable about the market and the competition.

Corporate moguls or line staff cannot put together a logical strategic or tactical marketing plan unless there is a firm foundation upon which to build. We did not yet have a firm market base for our Omega product. Our marketing plan, therefore, was more a plan of action, to help us rank our work and travel. It covered the entire spectrum of possibilities within both the commercial and military sectors and encompassed Japan, North and South America, and

Europe. Although we were covering the commercial airline market, we still believed that our big break would come from the military.

It was a frenetic period for all involved. Despite the slow start, we all remained committed to the product, including the vice president of the Avionics division, Kieth Glegg. This long-term belief in the product, the commitment and our ability to remain with it, would prove to be our salvation. It was a tough decision to stay the course. It meant a continuous investment in the product line. A critical decision for any company is knowing when to quit with a particular product. Not all products succeed.

There is always a temptation to continue with the investment for just a little longer in the hope that the situation will change. There is no magic formula and no panacea. The company must base its decision on an honest appraisal of all aspects of the situation. We never seriously considered dropping out of the Omega market, but achieving a return on our investment was going to take longer than we expected. It even appeared as if the advances in technology were about to overtake us, which meant we might have to start the design of a second-generation Omega product before the first one had captured a market.

Our strengths were that we had proved our ability to produce a technically sound product and that we were gaining a worldwide reputation as one of the two pioneers of airborne Omega Navigation Systems.

We entered 1974 in something of a disarray, with no focused marketing plan, but with a worldwide, far-ranging, and ambitious one. And we were still asking ourselves where do we go from here.

Early in 1974 the Canadian Armed Forces issued the long-awaited Request for a Proposal. It had decided to use Omega equipment in the Argus and other aircraft, in anticipation of the LORAN-A chain closing down. Although it invited Northrop, then the only other supplier, to bid, Canadian Marconi had a significant advantage. The U.S. Navy had paid for the development of the Northrop system, but the Canadian government had shared in the development cost of the CMA-719.

We were not complacent. We were not about to see our much needed launch customer slip away, and so we worked long and hard to complete that final phase of the procurement cycle. Finally, in the spring the Canadian Armed Forces rewarded us for our efforts with an order for almost 100 systems.

Over the next few years it ordered more systems, bringing the total sold to the Canadian Armed Forces to more than 150 ship sets. The Canadian forces installed those systems into several different types of aircraft. First, it fitted the system into the aging Argus maritime patrol aircraft and later its replacement the Lockheed P-3 Orion. It installed systems in its B-707, Lockheed C-130, and De Havilland Buffalo transport aircraft fleets and others.

The Canadian forces even installed systems in two submarines and several helicopters. Robert Baillie, our expert on unusual installations, had the exciting experience, or terrifying experience—depending on how you view excitement—of spending many hours in a World War II class submarine, not in the relative safety of Halifax harbor but underwater, in the middle of the frigid North Atlantic Ocean. He helped the Canadian forces design the installation, install the system, and then check its performance.

The contract with the Canadian Armed Forces was a modest beginning, but a beginning nonetheless. Under normal circumstances that contract would have launched the product. However, circumstances were not normal. We knew that, with the recent advances in component technology, smaller, lighter, lower-cost Omega navigation equipment was rapidly becoming possible. Market factors were also stopping us from fully capitalizing on the contract with the Canadian forces. Without that contract we might have abandoned further Omega development, and so in that respect, at least, it came at a crucial time.

In other respects, the contract came at the beginning of the end of an era. It marked the end of the Omega pioneering era, an era in which we had played a major role. It was not quite over, but it would be within a year. We, and others, had succeeded in convincing many in the aviation community to continue to consider the Omega System as a primary navigation service. The U.S. Coast Guard had overcome many obstacles to progress in the construction of the Omega stations. Those of us involved in the pioneering work had convinced the regulatory agencies to consider Omega equipment as an alternative to inertial-gyro navigation equipment. We had overcome much early skepticism and opposition, and we had steadfastly opposed the use of the Navy communication signals for navigation.

In retrospect, we had been naive in thinking that the Omega stations would be built on time and that the market would readily embrace this new navigation system. However, we were a medium-sized company operating in the big league. To succeed, we knew that we had to establish our position early, before the large avionics companies dominated the market. Innovation and the ability to identify specialized products were our hallmark. We expected, or at least hoped, that those innovative and specialized products would find a large market, as had happened with the Canadian Marconi Company's Doppler Navigation Systems. It was obvious that several of the large multiproduct avionics companies were watching the Omega market with the view to entering the fray. We were particularly concerned that large U.S. companies, such as Collins Radio, the Bendix Corporation, the Sperry Corporation, and Litton Industries, would enter the market before we had firmly established ourselves.

During the winter of 1974 I made my second trip to Brazil. On that occasion I carried the 35-mm projector, which was becoming a permanent fixture in my luggage, and more than 100 pounds of CMA-719 demonstration equipment. Once again, upon my arrival in Rio de Janeiro, Antonio Miranda worked wonders with the Brazilian customs officials. My five pieces of luggage and I got through customs just in time for Antonio and me to catch a domestic flight to

Porto Alegre, in the southern part of Brazil, where the principal Varig engineering and maintenance base is located.

Since my first visit six months earlier, Antonio had maintained contact with many people at Varig. In particular, he had kept Assis Arantes up to date on the progress of the Omega System. Varig appeared to be inclined towards using Omega equipment, rather than the expensive Inertial Navigation Systems, as a replacement for the LORAN-A equipment in its B-707s.

Our marketing was far from over. Varig would still need more persuasion, including proof that the system would work throughout its route structure. First, Varig wanted a bench demonstration. If that was successful, it wanted to borrow a system so that the airline could conduct a flight evaluation.

The Varig pilots and some of its engineers still wanted Inertial Navigation Systems in the B-707s. They knew those systems worked, and they remained skeptical about Omega. Also, the regulatory agencies still had not approved the use of Omega equipment as the primary means of navigation. Until they granted that type of approval, aircraft operators could only use Omega equipment as a backup to another system, such as a Doppler Navigation System. Varig, like most airlines with that generation of aircraft, had both Doppler and LORAN-A systems installed.

The CMA-719 system worked well during my bench demonstration in Porto Alegre, although the skeptics wanted to see the system receive real, not simulated, signals. I had brought an antenna with me. It took half a day to run a cable from the conference room to the parking lot and to find a suitable location for the antenna. I was not worried about cars running over the antenna. What I had to find was a location where the antenna did not pick up local electrical interference. Finally, the system received enough signals to prove that it worked in a real signal environment.

If only the skeptics had known how precarious that demonstration had been. We were in a part of the world where the Omega signal coverage was at the ragged edge. According to the laws of low-frequency signal propagation, and with only four of the eight planned Omega stations on the air, there should not have been enough signals for the system to work. However, luck and an anomaly in the ionosphere were on my side. We repeated the demonstration, later in the week, to more Varig pilots in Rio de Janeiro.

The Varig personnel were not the only ones who wanted to see the system process signals and to see proof that real signals were available. Once, when we were trying to win a contract with the Italian military aircraft manufacturer Aeritalia, now Alenia, its engineers insisted that I connect the system with an antenna. For a whole day I tried to get the system to work, but there was too much local electrical interference on the ground. Eventually, although it was raining hard, we took all the equipment onto the roof of a building. I wrapped the boxes in plastic bags to protect them from the downpour, and we all huddled under umbrellas while I demonstrated the system. The system worked, but we did not win the contract. It was awarded to the Northrop Corporation. That was a classic example of an early major order giving a product credibility. The Italians selected the Northrop system not only because of its technical merits but also because the U.S. Navy was using it. That was a better marketing story than our contract with the Canadian Armed Forces.

Porto Alegre is located in the province of Rio Grande do Sul. That is gaucho or cowboy country. While in Porto Alegre, Assis Arantes introduced me to one of the culinary delights of my life. At a restaurant called the Rancho Alegre, I was treated to a typical Brazilian churrascarai or barbecue. I ask vegetarian readers to try and find it in themselves to forgive me for my raptures over a Brazilian churrascarai.

At the Rancho Alegre, the waiters bring large pieces of meat to the table on a skewer the size of a medieval fighting sword. The waiter then proceeds to carve off slices of meat. Brazilian beef, lamb, and pork, raised on free ranges and uniquely barbecued, are the tastiest meats I have eaten anywhere in the world. I very quickly became a fanatic of Brazilian, especially southern Brazilian, churrascarai. My Varig colleagues were obviously delighted with my enthusiasm for their food. During my many return trips to Porto Alegre, I did not need to be asked twice when someone suggested that we go to the Rancho Alegre for lunch or dinner.

My biggest regret, in all my trips to Brazil, is that I did not extend my visits to see more of that beautiful country. In 1978 I did spend a week in Manaus, a small town in the Amazon, where the Rio Negro flows into the mighty Amazon River. The occasion was a conference, hosted by Varig, and attended by representatives of all the South American airlines.

Business travelers are rarely able to find the time—perhaps it is more accurate to say rarely take the time—to see a country they are visiting. We all have the habit of flying in, doing our business, and flying out again as quickly as possible. We see airports, hotel rooms, our customer's office, and the roads connecting them. Our time is valuable, or so we think. We always want to make the best use of our time. However, are we really making the best use of our time?

Over the years such trips take their toll, and burnout or just plain tiredness occurs. As a result, our companies no longer have the benefit of our hard won experience. Other, less experienced people now have to take over where we can no longer continue. I believe that, in the long term, we would be better off if we took advantage of the opportunities that international business travel can provide. Most of the time extending a trip by an extra day or two would not have a harmful effect on either our business or our home life.

Instead, we rush off to exotic places like Brazil. We spend 12 hours, flying overnight, to reach our destination. We go straight to work, although we have not had a decent night's sleep. We conduct our business during two or three days, and then we rush back, exhausted. On Saturday we are cutting the grass like zombies, half asleep, when our neighbor sticks his head over the fence to

Teaming a Product and a Global Market

tell us how lucky we are to visit such wonderful places. And our spouses complain because we are too tired to take the children to the park.

That second trip to Brazil was successful because Varig asked to borrow a system to test in a B-707. Back home we scrambled to find a spare system. In those early days of trying to establish a market, it was always a challenge to find a spare system to loan to a prospective customer. We had only a few systems available for that purpose, but the number of evaluations to which we were committing was increasing weekly. At one point I had a large card rack on the wall of my office. That was our inventory and flight test control board. With colored cards we tracked the equipment that was out on loan, the flight evaluations in progress, and our future commitments. Although we were building computerized equipment, we did not have, at that stage, personal computers in our offices.

Our Field Engineering Department helped Varig design an installation. Ormee Chamberlain journeyed to Brazil to survey the aircraft and to find a suitable location on the fuselage for the system's antenna. He went back in late June with Pierre Fournier of our Training Department. Ormee was to help Varig technicians install the equipment in the aircraft, and Pierre was to teach the flight crews how to operate the equipment. Afterward they checked the system during a flight across the South Atlantic. Varig then flew the system on regularly scheduled passenger flights between Rio de Janeiro, New York, Tokyo, and Lisbon.

The flight test program lasted for three months. Despite the marginal signal coverage, the system performed quite well and provided, on average, navigation to within 3 nautical miles. However, some unexplained signal anomalies produced much larger errors. Despite those problems, which Varig knew we would solve, it concluded the system satisfied its requirements.

Varig did not buy systems immediately. It decided instead to wait until the United States announced a firm date for the shutdown of the LORAN-A navigation service. That judicious wait saved Varig a lot of money because in the meantime we designed a second-generation, lower-cost system. We remained hopeful, at least during the following 12 months, that Varig would buy CMA-719 Omega Systems.

That first evaluation with Varig was the beginning of a continuing business relationship that helped to open the entire Brazilian aviation market for Canadian Marconi. For the time being, though, we had to be content with knowing that our system had been successful on the test flights.

The world's aviation community was finally waking up to the potential of the Omega System. We identified many potential customers. Most of them, of course, wanted to try a system before they would consider placing an order. We installed a system on a Belgian Air Force Lockheed C-130 cargo aircraft. The Ministry of Defense of France had its own aircraft test center, and it installed a system in one of its Caravelle test aircraft. The system remained there for more than a year. We were prepared to support such a long test program because we felt it would help us to penetrate the European, especially the French, markets.

We underestimated the extent of protectionism that was taking hold in Europe, with France in the lead. Eventually, the French government selected the French company Crouzet, working with another French company, Sercel, to develop an indigenous airborne Omega Navigation System. Several years later the Ministry of Defense bought a few systems from Crouzet, but the company did not succeed in parlaying that purchase into a significant market. Sercel, though, produced and successfully sold Omega equipment for ships.

For a time, Crouzet was a competitor that we had to watch carefully. We never succeeded with the French military, but later we did secure a significant and prestigious contract with Air France.

During 1974 we had the greatest difficulty scheduling our few systems for flight tests. Canada's national flag carrier, Air Canada, tested the system for the second time in a DC-8. Sterling Airways in England conducted its third evaluation of the CMA-719 in a Boeing B-727. That was the first time that we had flown the system in a B-727. We agreed to the evaluation to get B-727 experience because we were still trying to interest Boeing in the system. Naturally, we hoped that Sterling Airways would buy some systems. They never did. Laker Airways, the rapidly growing British charter airline that was pioneering low-cost air travel, tested the system in a B-707.

Pacific Western Airlines, operating out of Vancouver, used a transport aircraft in quite an unusual role. They had altered a B-707 to carry cattle and sometimes horses. Breeders contracted with Pacific Western to transport the livestock. Because the airline flew to destinations throughout the world, but mostly across the North Atlantic, Pacific Western needed a worldwide navigation system. It had been using Doppler Navigation Systems, built by Marconi Avionics in the United Kingdom, which were serving it well, but Omega equipment held the promise of greater accuracy.

Pacific Western borrowed one of our scarce Omega Systems. It was another job for our unusual installation expert, Robert Baillie. Between March and September, Robert periodically flew with the cattle. To check the cattle, which were in pens in the main cabin during the flight, Pacific Western Airlines engineers had installed an overhead trolley. A crew member would lie on the trolley and pull himself along the entire length of the aircraft.

Occasionally, the captain asked Robert to help out with the cattle check. He became a part-time cattle hand as well as the Omega flight test engineer. Such was the nature of the business; everyone pitched in when needed. The Marconi Doppler was serving the airline perhaps too well. In the end it did not buy Omega Systems. Pacific Western Airlines eventually retired the B-707s, and its new aircraft came equipped with Inertial Navigation Systems.

During that same year I made my first trip to Japan. Canadian Marconi retained the Okura Trading company, in Tokyo, as its representative for business in Japan.

Our contact there was Sam Takeuchi. Sam was not his Japanese first name; it was the name he had adopted because it was more easily pronounced by anglophones. Sam, cheerful, industrious, and professional, worked closely with several of us at Canadian Marconi for many years. He was instrumental in helping us to set up a business base in Japan for several of our products.

The primary purpose of my first visit was to set up a relationship with Japan Airlines, a likely Omega customer. Also, a contract with Japan Airlines would help our other marketing efforts in the region. During the long flight to Tokyo, I read, for the second time, *The Businessman's Guide to Japan and Its Customs*. I felt that I was well prepared. However, I was not prepared for the earthquake that shook the Japan Airlines building during our first meeting.

It was my first earthquake. I was terrified. While I hid under the conference table, with the room swaying and the lights flickering, my Japanese colleagues remained in their seats smiling. They were far too polite to comment on my cowardice. Apparently, for them, such events were a common occurrence, and they knew that the buildings were designed to withstand such tremors. The quake probably only lasted about two minutes, though it seemed much longer. Eventually, I regained my composure and was able to finish the Omega presentation.

The Japan Airlines engineers gave no sign of their interest in Omega. I did not expect any feedback at that first meeting. I knew from my preparatory reading that the Japanese do not react instantaneously, or even comment, in such a business situation. I knew that they would first read and analyze the product literature that we had given them. Then they would seek a consensus with their colleagues on their next course of action. However, the prospects appeared promising, at least to this optimistic "marketeer."

An important evaluation during that period occurred in a U.S. Air Force aircraft whose call sign was *Speckled Trout*. That aircraft was used to transport military officials around the world and served as an unofficial Air Force test aircraft for new avionics systems. We agreed to loan equipment to the Air Force because we had our sights set firmly on it as a prime potential customer. The CMA-719 system remained on board *Speckled Trout* for almost 10 years. During that time it provided both the Air Force and us with invaluable flight test data on the system's performance around the world.

We were trying everything we could think of to open the Omega market. We even investigated an entirely new, for Canadian Marconi, segment of the aviation market: business aviation. That market includes corporations that have their own aircraft, which are larger than the aircraft flown by sport fliers. The Beechcraft King Air is one of the smaller corporate airplanes. It holds up to eight passengers and is powered by two turboprop engines. The Canadair Challenger is one of the largest business airplanes. It is powered by two turbofan engines and can accommodate 20 passengers. Other airplanes include the turbofan-driven Cessna Citation and France's Dassault Falcon. A few corporations use larger commercial transport aircraft, but they are the exceptions in business aviation.

Those aircraft are usually for the exclusive use of the owners or chief executive officers of companies and their immediate staff or family. Many executive jet aircraft can cross the continent or the Atlantic Ocean without making intermediate stops for fuel. In keeping with the luxurious interiors, the owners of those aircraft frequently also want the latest in aircraft equipment. Several of the owners are also aviation buffs.

During the 1970s the business aviation market was huge. Business aircraft manufacturers delivered more than 5500 aircraft during the peak year of 1979. Then corporations started to view business aircraft as a luxury, and orders declined cataclysmically. Only 1000 aircraft were delivered in 1981. In the 1970s, however, the business aviation market was larger than the military transport and commercial airline markets combined. It was a market that Canadian Marconi could not ignore.

The business aviation business environment is quite different from either the commercial airline or the military aviation markets. Business aircraft coming off the assembly line usually do not have any interior fittings and have only a minimal amount of aircraft systems installed. A second party customizes each aircraft for its eventual owner. A modification center, called a fixed base operator, does the special outfitting. There are many fixed base operators throughout North America and Europe, as well as a few in Asia and South America.

In business aviation, avionics suppliers have to sell to the aircraft manufacturers and the fixed base operators, as well as the individual end customers. Also, the suppliers have to sell each system, one at a time, unless they are fortunate enough to persuade the aircraft manufacturers to make the system part of the basic aircraft. That is very difficult to do because most aircraft manufacturers want to give their customers as much flexibility as possible.

The whole approach to the business aviation market was radically different from anything that we had tackled before. We entered the market with some trepidation. Marketing was bad enough, but there was also the larger problem of supporting the equipment once we had sold it. Business aviation customers, unlike a commercial airline or a military organization, do not have spare equipment or even technicians located around the world. If the equipment breaks down, they rely on a local fixed base operator or the equipment manufacturer to fix the problem—not within days but within hours—and they could be anywhere in the world. However, we could not ignore the market because it was so large. Understanding it was a learning experience for all of us.

Within our Avionics Product Support Group two individuals had the courage to pursue the business aviation market. Carl Hanes and Tom Kelly, both Marconi veterans, single-handedly launched our entry into that market. Kieth Glegg supported the endeavor, even though I suspect he did not believe they would be completely successful. To his credit, he was prepared to let them find out if the market was suitable for a company like Canadian Marconi, which was more used to dealing with large customers, not myriad small ones.

Carl had been with the company all of his working life and was about 50 years old. Tom Kelly, an Irishman about five years older than Carl, had come to Marconi from Ireland's commercial airline, Aer Lingus.

One might wonder why two individuals from the Avionics Product Support Group were dabbling in marketing. Another of Canadian Marconi's strengths, in my opinion, is that its management does not confine an employee to a rigid job description. The company encourages any person with an idea for improving the business to bring it forward. If the idea meets with general acceptance, most likely the proposer will also be encouraged to participate in implementing the idea. It is the concept of the product champion applied across the company and to a variety of disciplines.

Carl and Tom started their odyssey. They visited business aircraft manufacturers and modification centers, as well as the regulatory agencies in Canada, the United States, and Europe. They visited several of the larger corporations that owned business aircraft. They learned about that segment of the aviation market, how to sell to it and what it wanted. They listened while the aircraft operators told them how they expected the company to support its equipment. They even sold a few CMA-719 systems. It was a cautious beginning, but a beginning.

One of the first fixed base operators to install the Marconi Omega equipment was a company called Butler Aviation located in Chicago. Over the years, their avionics manager, Tom Kokozinski, continuously provided us with encouragement and help. He helped us to understand the business aviation market and to overcome the barriers to entering that market. It was a classic case of two companies respecting and helping each other because the employees came to respect and trust each other as individuals. We were not just two organizations doing business.

Our relationship with Tom Kokozinski is one more example of a friendship developing between a customer and several employees at Canadian Marconi. I remain convinced that personal relationships, built on mutual trust, respect, and a willingness to help each other are the mainstays of the business world. We often forget the all important personal aspect of our business dealings. We all would be better off, and more profitable, if we reminded ourselves, from time to time, that companies are not the bricks and mortar of the buildings in which we work but the people with whom we work.

7

A Second-Generation Omega Navigation System

*Don't go around saying the world owes you a living;
the world owes you nothing; it was here first.*
—Mark Twain

From a marketing perspective 1974 had been a busy year. At least, we had done a lot of traveling and the preparatory work to establish our reputation in the international military and civilian Omega markets. We were anxious for those markets to unfold. We had our launch customer, the Canadian Armed Forces, and it was about to start installing CMA-719 equipment in several different types of aircraft.

Our engineers had completed the design of the CMA-719. What remained was sustaining engineering, which is the ongoing work to solve production or customer problems. Sustaining engineering also includes the design of new equipment features in response to customers' requests. There was much sustaining engineering work to be done; however, most engineers do not relish that kind of work. They prefer the challenge associated with the research and development of new products. Our Omega engineers would have become restless except for the most important event for the Omega Lab in 1974.

We realized that to remain competitive we had to start designing a new system, and so early in 1974 we began work on an airborne Omega Navigation System that would become known as the CMA-734. Improvements in avionics technology were making the CMA-719 obsolete, and already the systems that were using the U.S. Navy's communication signals for navigation were smaller and soon would be less expensive. Several U.S. companies, such as the Bendix Corporation, Litton Industries' Amecon Division, Tracor Incorporated, and the small company Dynell Electronics, declared that they intended to start the full-scale development of airborne Omega Systems. The competition was going to be tough, that much was obvious. The only way Canadian Marconi could remain competitive, with such a formidable group of companies, was to have a lower-cost product that performed better than the competition.

Northrop faced the same dilemma. It started to enter the second-generation Omega equipment market, then dropped out of further competition, even though it had been successful in selling many of its first-generation system, the AN/ARN-99, to the U.S. Navy, the Belgian Air Force, and others.

With only a few exceptions, the original Canadian Marconi Omega engineering team was intact. By this time, Don Mactaggart had gone off to work on other new products. Michel Galipeau was still the product manager, and Peter Gasser had now become his senior project engineer. The project engineer in Canadian Marconi's Avionics Division leads the day-to-day engineering activities of the engineering team. Strategic technical decisions are made with the product manager, but most of the time the project engineer runs the show. That is in keeping with the company's philosophy of granting authority as low as possible in the organization. As other successful companies have experienced, that approach to management challenges, motivates, and nurtures the talents of all employees.

The decision to invest in the design of a second-generation product before we had recovered our investment in the first one was a courageous but necessary decision. We had received some financial help from the Canadian government for the development of the CMA-719, and we hoped it would support the CMA-734.

Peter Gasser led the design activities for the CMA-734. The engineering team included many innovators. Peter, an experienced engineer, was never afraid to try something new. Joe Mounayer was still our electronics equipment packaging wizard. Jean-Claude Lanoue was our expert in both software and advanced circuit designs. David Bailey, a recent engineering graduate but a solid digital circuit design engineer, was on the team. We still had Ron Miller, an experienced and multitalented electronics engineer, and Gilbert Boileau, now the backbone of the Omega software team. Robert Baillie, who was becoming recognized throughout the world as an expert in the vagaries of Omega installations and the real world Omega signal environment, started work on the design of several innovative Omega antennas.

The team's primary goal was to produce a low-cost system. However, we agreed that they should not achieve a low-cost design by jeopardizing the performance that we had already demonstrated with our first airborne Omega Navigation System.

The marine Omega receivers required the operator to manipulate several controls to obtain a position fix from the equipment. That approach was too cumbersome for an aircraft environment, although within a year Dynell tried to convince the aviation community that the lower cost of those systems outweighed the inconvenience of their manual operation. It did not win the argument, but only after those of us producing automatic airborne Omega equipment launched a concerted counteroffensive.

The automatic Omega receivers are those systems that, with very little operator control, automatically calculate their geographical position. Their most important feature is an ionosphere variation software program that corrects the received Omega signals used to calculate navigation information.

The general aviation market was booming now, and that was our primary target market for the CMA-734. Nonetheless, we could not rely solely on succeeding in that market. We reasoned that the system we produced also should satisfy the requirements of helicopter operators. The fixed-wing aircraft fleets of military and commercial airlines were not our primary market for our new system. However, we wanted a design that we could adapt later for those markets.

We were asking the engineering team to produce a low-cost, small, lightweight automatic airborne Omega Navigation System that had at least the same navigation performance and operational features as the first-generation systems. Also, we did not want to discount any market opportunity. The request represented the ultimate challenge for any electronics engineer.

The team produced a product that was a third of the cost of our first one, half its size, and one third its weight and that consumed the power of an average domestic light bulb. More remarkable was that Peter's team made those reductions without jeopardizing the performance of the equipment. In fact, it navigated better than its predecessor and was more reliable, mainly because it had a modern processor and memory. Another remarkable aspect of the design activity was that within 18 months our engineers were flight testing a prototype, and production started within little more than two years after we began the design work.

Our customers loved the technological environment that was evolving. It was spawning greater competition in many areas. Now they had a chance to challenge the industry.

Manufacturing companies can lower costs by producing large quantities of a product. However, the avionics market, with only a few exceptions, compared with the consumer market, is small. Usually our production runs are for batches of hundreds of units, not thousands. We could not rely on large production runs to lower our costs. Instead, our engineering team had to discover novel and economical circuit designs, uncomplicated packaging techniques and a design that our manufacturing department could produce efficiently. The team members rose to the challenge.

In every office at Canadian Marconi there is a black or white board. Any day you can find a group of engineers clustered in front of a board discussing and solving a technical problem. Such problems are not given to a group of distant technocrats to solve; the engineers execute their own solutions.

What evolved was a novel approach to avionics packaging and an efficient Omega Navigation System design. Three years later, the adaptable packaging concept that we had adopted helped us with a critical delivery situation.

The CMA-734 had three units just like its predecessor, its successors, and those systems of the other Omega equipment manufacturers. Some aspects of the design just could not change. The system had an externally mounted antenna, a pilot's control and display unit, and a receiver-processor unit.

Robert Baillie designed several versions of the externally mounted antenna for different aircraft. We were one of the few companies to recognize that need, and those antennas demonstrated our willingness to respond to the specific needs of our customers.

Outwardly, the control and display unit did not change dramatically for the first few years because pilots favored a standard format. Its internal electronics, however, did change dramatically when we found more efficient technologies.

The largest technical changes occurred within the receiver-processor unit. Our engineering team continuously strived to develop the most modern designs. Jean-Claude Lanoue even tried to push the state of the technological art to its limits, and he often succeeded.

The circuit cards in avionics are usually mounted laterally in what is a rectangular-shaped box. Those cards plug into connectors mounted at the bottom of the box, much like files in a filing cabinet. That design, although conventional, has some disadvantages. When a file is in a drawer, its contents cannot be seen. Similarly, an electronics circuit card mounted conventionally cannot be seen. It must be taken out of the box or put on an extension lead to make a repair. Another disadvantage of that approach is that the size of the circuit cards is limited to the short dimension of the box. That means many cards with connectors and wires interconnecting them, all of which increase the electronic component parts and production costs. The problem of excess wires is usually overcome by using another circuit board with the wires printed on the board and connectors mounted on it. The circuit cards plug into the motherboard, which is in the bottom of the unit. Motherboards, however, are expensive.

The packaging design that Peter and the engineering team dreamed up on their blackboards was novel and unconventional but low-cost and more maintainable. It was a bold and daring decision. We were operating in a conservative market where customers did not always welcome changes to convention, technology excepted. Our engineers concluded that they could mount the electronics on just four circuit cards, provided they used cards that they could mount longitudinally in the box.

We had come a long way since the CMA-719 with its 22 circuit cards. The design also avoided the expense of a printed circuit motherboard and made access to the circuit cards easier. With that design, a technician could open the unit like a book. No one, to my knowledge, had previously tried such an approach with an avionics product.

The electronics design itself used the latest in electronic component technology. Many avionics products use several specially designed parts. Those parts are often expensive, which means the source of supply is limited. Peter challenged the engineers to design the product with no or, at least, very few specialized parts. Except for special crystal filters, the engineers met the challenge. The rest of the product used standard industrial parts, which were available from several sources. That decision also helped to reduce the cost of the system.

Our experiences with the first Omega design gave the engineers the background and confidence they needed to make significant improvements in the design of the receiver. A vital part of the system, the receiver has to find the Omega signals among many other frequencies and electrical noise and then convert those signals into a format that can be used by the system's computer.

The computer, its memory, and software were all new designs. The computer corrects the signals with the help of a complex software program. The software program eliminates the errors that the signals pick up while traveling halfway around the world and calculates the position of the aircraft. Also, it combines that information with other data to calculate all of the navigation information for the pilot.

Another significant improvement in the CMA-734 over the CMA-719, its predecessor, was its memory. We no longer had to rely on ferrite and magnetism to store the software. Instead, random access memory and read only memory circuits were available. Not only were those circuits much smaller than the earlier memories, but also they were much more reliable.

Surprisingly, the memory capacity, at least initially, was the same as that used in the first-generation system. The engineers thought that 8000 words of software would remain adequate for the complex Omega signal processing algorithms and the navigation software. We did not foresee a need for additional memory capacity. We also did not envisage our customers' insatiable appetite for more features. To satisfy both needs more software was needed. Fortunately, the engineers had designed the system so that they could change the software and add memory and other circuits later.

This was a period of rapid technological change. However, for the time being, our engineers once again had created a product design that was ahead of the competition. Our engineering team worked with a management team that both encouraged creativity and had the courage to innovate.

8

Looking Back

You can't build a reputation on what you're going to do.
—Henry Ford

I have not told the entire story of the Omega pioneering era. Instead, I have waited to describe some events. In many ways, these events sealed the fate of this era for the Canadian Marconi Company. It was a time for reflection. More important, it was a time to decide how we should tackle the future.

We continuously had to deal with three factors that frustrated our progress. We had no control over the construction of the Omega ground stations. Their delay was causing most prospective customers to wait or to buy Inertial Navigation Systems. Canadian Marconi, although a moderately large company by Canadian standards, was small when compared with several of our competitors. Also, we were not as well known in the global market. The fact that we were not a U.S. company with a strong domestic market to build on just made our job more difficult.

Looking back, it appears as if the pioneering era closed rapidly, and with its closing our situation changed. In fact, events unfolded during 18 months between 1974 and 1976, which was a period of frenetic activity and challenge. Our challenges came in many forms. One, of course, was the development of a second-generation airborne Omega Navigation System, the CMA-734. The market prospects remained promising. However, it was obviously not going to be an easy market to penetrate or to dominate.

During this period the U.S. Air Force announced that it was going to start a major Omega equipment procurement activity. We had waited a long time for such an announcement. The FAA's proposed plan to close down the LORAN-A navigation chain had prompted the Air Force into action. Many of its long-range transport aircraft relied on the LORAN-A system, and so it would have to replace the receivers.

The U.S. Air Force, aware of the recent advances in technology, challenged the avionics manufacturers to develop a low-cost system. It went so far as to declare that it expected to buy equipment for less than $15,000. That was a major challenge. The first-generation systems had cost about $60,000.

Initially, the U.S. Army and Navy joined the Air Force in this procurement activity. A triservice procurement was most unusual. The prize for the winner would be enormous. The Air Force launched Project 2041, the name it gave to the procurement activity. Soon after, the Navy and the Army dropped out. Nevertheless, we expected the Air Force to buy more than 1000 systems.

The Air Force had flown the first-generation systems built by Canadian Marconi and Northrop. Marconi and Northrop were the undisputed pioneers, and so we naturally considered ourselves to be the favorites.

By now, Tracor Incorporated, the Bendix Corporation, and Dynell Electronics were well advanced in their designs of airborne Omega Systems. Despite those competitors, we considered Northrop to be the company that we would have to beat.

The Air Force decided to release its technical specification before releasing the formal Request for a Technical and Price Proposal. It planned to conduct a technical evaluation of the competing designs in a laboratory. After that evaluation, it intended to select three companies for the final phase of the competition. To add to the drama, the Air Force selected a five-day period during which the competing companies would have to show that their equipment performed to the Air Force specification.

The U.S. military usually selects a supplier on the basis that it is the lowest cost compliant bidder—four unambiguous words and an equally unambiguous statement of intent. The problem with that method of procurement is that it assumes the customer's specification is realistic.

The U.S. Air Force was aware that it would be some time before all of the Omega ground stations would be operating. That meant the signals from the stations that were working would be weak in parts of the world. To counteract that weakness, the Air Force wrote into its specification a requirement that the airborne Omega receiver be capable of finding extremely weak signals among the noise.

The problem was that the threshold the Air Force chose for that signal-to-noise ratio was unrealistic. We tried, in vain, to convince the Air Force that the current technology would not allow for such a sensitive receiver. Northrop also tried to convince it to relax the specification. The Air Force ignored the two companies that had the most experience with airborne Omega equipment.

In our Omega Lab Peter Gasser and his team worked frantically to complete a prototype of the CMA-734 and have it ready for the demonstration in the Air Force laboratory. We also wrote our response to the Air Force Omega Navigation System specification. We were honest. We stated that we could not conform to the signal-to-noise ratio requirement. We offered a receiver sensitivity that was close to the requirement and realistic. Later, we found out that Northrop had responded in exactly the same fashion. The ratio was the only part of the specification with which we did not comply. The other three competitors stated that they could meet the requirement.

The day of the demonstration approached. Luck and solder held our prototype equipment together. The Air Force had scheduled our demonstration for a Monday, and so Peter and three other engineers left for Dayton, Ohio, on Saturday to give themselves ample time to be ready. They took with them soldering irons and a suitcase full of spare parts.

I remained in the Omega Lab, chewing my fingernails, hoping the telephone would not ring with someone requesting that I dispatch some unexpected spare part. It rang once, and a spare memory board was on its way by courier service to Dayton.

Peter and his team had the equipment set up in a hotel room all weekend. They nursed it to make sure it would work during the demonstration. We would only have one chance. The Air Force intended to drop any company that was not able to prove that its equipment conformed to the specification. Our system worked as we had said it would. The sensitivity question, though, still hung like a dark cloud overhead.

A few months later the Air Force cut Northrop and Canadian Marconi from the competition. It had judged that our systems were not sensitive enough. It was unbelievable that the two companies with the most experience were out of the competition. Somehow the other companies had convinced the Air Force that their systems would meet the specification.

During the next 12 months we spent most of our time at the Pentagon, Wright–Patterson Air Force Base, and the Military Airlift Command at Scott Air Force Base. Repeatedly, we tried to convince the Air Force that our system would work better than any of the three it was then considering.

We offered the Air Force a solution. Because Dr. Frugenfeld had shown that the Navy's very low frequency communication signals could be used for navigation, we offered to add a similar receiver to our equipment. Our plan was to use the communication signals to augment the Omega signals.

The solution was Kieth Glegg's idea, and he tasked Jean-Claude Lanoue with the job of designing such a receiver. He even held weekly meetings with the principal engineers to hear about their progress and was often in the Omega Lab discussing the design with Jean-Claude. The Air Force did not change its specification or accept our suggestion. However, we offered the new receiver as an option to all our customers, and that helped our sales.

The Air Force continued its competition and two years later awarded a contract to Dynell. Initially, the system had problems receiving weak Omega signals, and many changes to the equipment became necessary. We felt vindicated, but we did not have the Air Force business. Eventually, another company bought Dynell, and a short time later it dropped out of the Omega business.

A few years after the Air Force competition, at a cocktail party, I spoke with a retired Air Force officer. I asked him what we should have done in the situation that we had faced. He replied with one word, "Lied."

Our failure to win the Air Force competition was a very serious setback. We had a difficult decision to make. Should we continue in the Omega business? Northrop looked as if it would soon drop out. Litton, Bendix, Tracor, and Dynell looked as if they would be serious competitors. The French company Crouzet was developing an airborne Omega Navigation System.

Competing for many of the same customers were Global Navigation Systems Incorporated and the Communications Components Corporation. However, the regulatory agencies, the U.S. Radio Technical Commission for Aeronautics, and potential customers were still hotly debating the appropriateness of using the communication signals for navigation. The debate continued for several more years.

We had started in the Omega business in 1968. Five years later our first product, the CMA-719, was ready to go into production. In 1974, we had started the design of a second-generation system, the CMA-734. However, we had only one significant customer, the Canadian Armed Forces. The U.S. Navy was using the Northrop equipment, and the U.S. Air Force was now looking elsewhere for an Omega supplier.

Should we have started in the Omega business so early? Should we have so readily accepted the mantle of Omega pioneer? Perhaps we should have paced our development to coincide better with the construction schedule for the Omega ground stations. We had invested considerable funds and effort to reach this point. What did we have to show for all our investment? Not much, at least where it counts—lots of revenue.

We could have put in place a small Omega research and development program. That would have allowed us to develop some of the key technologies we would require when it was more timely to launch a product development. It also would have allowed us to monitor and to assess the technical developments of the Omega System. Because the early 1970s were a low point on the company's revenue curve, such an approach might have appeared more reasonable. Looking back, I believe that Canadian Marconi's management made the right decisions and took the right course, even though initially the cost was high.

Between 1969 and 1972 the worldwide avionics business had been declining. Sales were 30 percent lower in 1972 than they had been in 1969, and significant growth would not occur until 1978.

In the 1960s, Canadian Marconi Avionics Division had been a one-product division. The company had sold many aircraft Doppler Navigation Systems, which remained a significant source of revenue. However, if the division was going to prosper and grow, it would need more products.

Kieth Glegg understood the need to develop new products. That this need occurred during a recessionary period was unfortunate. Canadian Marconi had several product champions, and they were not short of ideas.

One young engineer, Sol Rauch, concluded that it was possible to design aircraft instruments that did not have any moving mechanical parts. At that time, aircraft instruments were electromechanical devices, and their insides looked like a Swiss watch. Sol's concept replaced the clockworks with digital circuits. He proposed replacing the instrument's rotating pointer with a ring of light. The idea was to illuminate the ring from the zero point to the point on the dial that corresponded with the value the instrument was measuring. It was a novel and innovative approach.

With a small team of engineers, Sol first tested that concept by designing an altimeter for our Doppler systems. Later he had the idea that if he uncoiled the circular digital instrument, he would have a vertical instrument. Although vertical instruments were unheard of, Sol believed that the concept would substantially improve the way engine information was displayed.

While that work was going on, Don Mactaggart championed the airborne Omega Navigation System development project.

Another engineer was championing a third product idea, an aircraft navigation management system. Aircraft have many different navigation systems onboard, and each has a particular role to play. It was the job of the navigator or pilots to interpret the various pieces of information.

The company planned to design a navigation management system to take the place of a navigator and to reduce the pilot's workload. The proposed system would accept electrical signals from a variety of systems on an aircraft. It then would mix all those data and provide the pilot with navigation information. The Canadian Marconi system would even be able to tune the aircraft radios. The Marconi engineers also decided to display all the system's information on a cathode-ray tube in the cockpit. At the time their concept was a major break with traditional navigation practices.

Canadian Marconi was not the only company planning to develop a navigation management system, but the Marconi design was the most ambitious. Kieth Glegg approved the development of those three products. Only the navigation management system did not become a significant source of revenue for the company. It was an idea that was far ahead of its time.

This bold development program was expensive. Also, the markets for the engine instruments and the Omega System took longer to develop than expected. The instruments were novel. As a result, it took a long time to convince pilots to accept this radically different approach to the display of engine information. Eventually, the U.S. Army helicopter crews embraced the concept.

The questions still remain. Should we have waited until the Omega concept was more mature, and should we not have been a pioneer? As a foreign company—most of our sales were for export—we had to try harder.

We had to be in the forefront. Being one of the first companies in the business helped us to build our credibility. Companies have to earn such credentials. No company, large or small, can expect its customers to assume it will succeed with a new venture. Customers may feel more comfortable with a large, well-established corporation, believing its size alone will sustain its commitment. However, even some of the large companies that started in the Omega business eventually pulled out. Their reasons are known only to themselves. Perhaps they simply had other more profitable products.

The Canadian Marconi Company remained committed to the Omega System. We helped to educate the aviation community about the system. We were one of its principal advocates. We supported industry initiatives to write specifications controlling the installation of Omega Systems and their certification in aircraft. Our people traveled the world, marketing the system, our product, and our company.

Finally, we committed to the development of second-generation equipment, even though we had not recovered our investment on the first system. Even after we failed to win the large U.S. Air Force contract, we stayed with the business. No one could doubt our commitment. By the early 1970s, most people in the aviation community thought highly of our products and our engineering talent. They understood our failure with the Air Force and did not view it as a technical deficiency.

We had earned our credentials. We had become recognized throughout the world, and we were ready. Consequently, when the market materialized, no one doubted our ability to perform. I no longer had to respond to questions during marketing presentations such as: "Macaroni, wasn't he the Italian guy who invented the radio?"

Our prospective customers would have viewed us quite differently if we had simply waited and watched until it was safe to enter the Omega business. Our ability to continue investing in the product's development was possible due to some financial help from the Canadian government. This was a classic example of industry and government working productively together.

Despite the financial help from the Canadian government, our own investment was large. We paid the major share of the design costs. We were financing the startup costs for the manufacture of the equipment, and we were building equipment as an investment against possible sales. Also, we had to support customer tests and to pay our marketing costs.

Because the avionics industry was in a recessionary period, our costs were even harder to absorb. We could barely afford the investment in one new product. Nevertheless, we had three expensive projects and other smaller ones under way. Kieth Glegg had the foresight to understand our need for new products, even though the financial investment was considerable.

Corporate managers of successful companies know when a company is moving into a new phase in its evolution. They recognize the need for a different type of talent to guide the operation in the new phase. In 1976 Kieth Glegg left the Canadian Marconi Company. His task was done. He had helped to build the Avionics Division. He had introduced an effective organizational structure built around product managers. The product managers, by virtue of their talent, selection and position, were the champions. They kept a product alive.

Kieth had set the division on a course of aggressive new product development. After he left Canadian Marconi, he became vice president for industry at the National Research Council of Canada, where he continued to help Canadian companies strive for excellence.

We needed someone at the helm who could turn our new emerging systems into profitable business products. If we were to remain profitable, that new leader also would have to exercise fiscal and strategic restraint. It was not time to launch more development programs. We had to work out how to complete and to market efficiently those products we had under way. It also was time to evaluate, honestly, the prospects for our new products. We faced the prospect of canceling some programs.

The person chosen for the job was John Simons. He became general manager of the Avionics Division in November 1976 and vice president one year later.

John brought to the top job in the division a precise form of management. He was not as flamboyant as Kieth, and he was more serious by nature. John was also less inclined to take further risks.

John's degrees—a bachelor's of science in engineering and a master's in business administration, both from McGill University—served him well. His career had developed in the Avionics Division; he knew our products and understood our markets. He challenged our technical decisions. Mostly, though, he challenged our investment and bid price suggestions. He established stringent financial control measures and a more disciplined approach to project management. However, he did not change the product management organizational structure that Kieth Glegg had introduced.

John Simons' ascendancy to the top job in the Avionics Division brought the type of leadership we needed. Later, John became president and chief executive officer of the Canadian Marconi Company.

The change in leadership occurred almost two years after the point at which I left our story in the last chapter. However, the period between the development of our second-generation Omega Navigation System in 1974 and the change in leadership in 1976 was a transitional phase. It proved to be the end of the Omega pioneering era. It was also the beginning of the next phase, an era when, through a continuing commitment, we reaped the benefits of our search for excellence.

Part Two

A Global Market

9

"I'm From the Government and I'm Here to . . ."

*Look up and not down; look out and not in;
look forward and not back; and lend a hand.*
—Edward Everett Hale
(Chaplain, U.S. Senate, 1903-1909)

The title of this chapter probably produces reactions of pride, skepticism, or humor. Why those divergent reactions? It depends upon who you are and whether you are the one making or the one hearing the announcement.

The regulator feels pride, the object of the regulation may feel skepticism, and an observer may only see humor. For many, those reactions may be understandable—but they are not valid. We all are on the same side.

A government is not in the business of stifling growth. On the contrary, its job is to help a nation, its social structure, and its businesses to grow and prosper. Domestic revenues can maintain a country, but to grow a country must export its goods and services.

Industries, national and international trade organizations, trade unions, and yes, even international regulatory agencies and governments all must work together in a climate of respect, knowledge, and mutual trust. Only then can the players expect to prosper.

The avionics business is a complex and highly regulated industry. Consequently, many government agencies and trade associations around the world influence our business environment.

Over the years, the Canadian Marconi Company has invested considerable resources to play its part as a full member of the international aeronautical business. The company has reaped the benefits of that investment. We have also watched companies falter as they sat on the sidelines, expecting others to show the way. The adage "you only get out of something what you put in" is true in most of our business, academic, and social endeavors.

Although I am writing about our participation in the management of the international avionics business, our willingness to participate is a lesson that others should heed, especially those who now are embarking on a transition from the military to civilian markets.

It was a Canadian government program, with origins in the defense sector, that helped us to start in this business. As early as 1940, Prime Minister Mackenzie King and U.S. President Franklin Delano Roosevelt recognized that both countries would benefit if they coordinated the work of their defense companies.

One result was the Canada–U.S.A. Defense Production Sharing Arrangement, which was signed in 1959. The arrangement gives Canadian and U.S. companies the right to compete under the same conditions for opportunities in both defense markets.

To help compensate for the large research and development contracts that defense agencies of other nations give to their companies, which the companies can then use for commercial products, the Canadian government introduced an effective made-for-Canada program. That program was called the Defense Industry Productivity Program, or more commonly, DIPP.

Through DIPP, the Canadian government shared the development costs of a product that had a strong possibility of creating export, particularly defense, revenues for Canada. This is not subsidization by government, as some of our foreign competitors periodically tried to claim. Companies had to invest more than 50 percent in the design costs, and there were pay-back conditions. DIPP is one Canadian solution for redressing the imbalance caused by our smaller size. Other programs cover marketing in new territories and basic technological research; however, DIPP has been the most effective program.

Several of our products had their beginnings through a close collaboration with the Canadian government. The Omega product line is an example. The technological and market risks were too large for the company to proceed on its own. Also, our only competitor at the time, Northrop, had a fully funded development contract with the U.S. Navy.

The Canadian defense electronics industry on average reinvests 15 percent of its revenue in product research and development. The Canadian Marconi Company has routinely exceeded the industry average. It recognizes that it must make prudent reinvestments to achieve growth and prosperity.

Preparing the Omega DIPP proposals was always a challenge, which was often undertaken by Peter Gasser and me. Peter would write the development plan and I would concentrate on the marketing plan. I usually undertook the job on the dining room table at home on weekends. It is rare for anyone in a successful dynamic business to find enough quiet time at work, and successful teams understand this inescapable aspect of business.

Teaming a Product and a Global Market

Later, during our search for excellence, the government's industry, science, and technology official assigned to review our DIPP contract work was Peter Trau. Peter, an immigrant from Austria, is a technocrat with industry experience. Forthright in his opinions and a competent guardian of the taxpayer's purse, Peter was a person we enjoyed working with and respected.

On several occasions Peter would send us away to improve our ideas and plans—usually my marketing plan or market projections. Government and industry should not, indeed cannot, work in a confrontational environment. Peter's questions always prompted us to review what we wanted to do more objectively; he was our conscience.

The industrialist has an obligation to the company's stakeholders and the government employee to the taxpayer. Each works in a different environment with its own methods of checks and balances. Both parties need to understand each other's duties and work environment. We did, and I think that is why we succeeded. Successful companies understand that industry and government are on the same side, with the common goal to secure export revenues or to succeed with an international joint venture. We always tried to work with, not against, the Canadian government. People like Peter Trau helped to make our business more successful, and we owe him a debt of gratitude.

Before the airlines could use Omega equipment for navigation, the national regulatory agencies around the world would have to sanction it use, and we expected the FAA would take the lead. We had visited the FAA in Washington, DC, on many occasions. As its name implies, that agency is responsible for all aspects of air safety in the United States. Our stated purpose for the visits was to keep FAA personnel up to date on our progress and Omega equipment flight tests. Our unstated reason was to build their confidence in the Omega System.

Each country has its own aviation agency, which writes its own regulations to govern our industry, although the agencies try to coordinate their work, so that there is some uniformity in the industry. We kept in close contact with most of the national aviation regulatory agencies. Those visits were a burden on our already meager resources; however, we understood that if we were to succeed, we had to make that investment.

Most countries belong to the International Civil Aviation Authority, which has its headquarters in Montreal. The International Civil Aviation Authority controls air operations that affect all member states, especially oceanic air operations. For years the airlines had complained that because they had to keep aircraft at least 120 nautical miles apart, they could not always fly the most efficient routes across the North Atlantic.

In 1975, after much protracted debate, the International Civil Aviation Authority announced that starting on Jan. 1, 1978, new navigation standards would take effect on the North Atlantic. Those standards included a reduction in the separation of airplanes to 60 nautical miles, provided they had certain navigation equipment. Aircraft not suitably equipped would have to fly the inefficient northern route via Greenland.

Our big coup was that we succeeded in persuading the International Civil Aviation Authority, through our national delegations, to declare the Omega System a suitable navigation system for the new separation standards. Its only restriction was that, initially, the airlines would have to use Omega equipment with the proven Doppler Navigation Systems. Only the Inertial Navigation Systems were considered accurate and reliable enough to be used on their own.

At the same time, the U.S. government announced a firm decision to shut down the U.S.-sponsored LORAN-A ground-based navigation system. The shutdown date was also Jan. 1, 1978.

Those of us in the Omega community finally had our big break. It was 1975, and so we had three years to establish our market or fail. This type of opening only comes once in the evolution of a product or market. We had an enviable reputation throughout the prospective Omega market, but this was not the time to rely on our reputation alone—it never would be. To remain successful you always have to put forward your best effort; that much we had learned.

We also knew that because the international regulatory community now had endorsed, at least partially, the Omega System, other avionics manufacturers would enter the market. The large Bendix Corporation had already started the design of an airborne Omega Navigation System.

Litton, one of our competitors in the ill-fated U.S. Air Force Omega competition, had transferred its Omega project to its Aero Products Division in Canoga Park, California. That division was enjoying considerable success with its Inertial Navigation Systems, and so it had the market base that we lacked. Our biggest concern was that if Litton should enter the Omega market, it would be a very serious threat to our business prospects. We were right to be concerned, but we could do little about Litton's position in the marketplace then or in the future.

Tracor Incorporated continued with its Omega equipment design work. We did not know how to react to Dynell Electronics, the small company that eventually won the U.S. Air Force contract. Although the design technique Dynell was using was radical, it gave the company a less expensive product. As a result, it remained a threat too. Our pioneering partner, Northrop, was beginning to drop out of the Omega market, but we could not discount it entirely.

We had been working with the Omega System for so long that we were confident that one day we would prove that aircraft operators could use the system as the primary means of navigation. However, for the moment we had to be content in the knowledge that the regulatory agencies were sanctioning its use with a Doppler Navigation System.

Like most industries, the airlines avionics community has an association that deals with matters of mutual interest: the Airlines Electronic Engineering

Teaming a Product and a Global Market

Committee. Also, like most industry groups, it is an organization that suppliers must support, if they expect to succeed.

The principal purpose of the committee is to help the airlines assess their future avionics needs. Then, through various subcommittees, its members write documents known as ARINC Characteristics to reflect those needs.

Each aircraft has a common set of avionics; however, there may be several manufacturers for any one type of equipment. An ARINC Characteristic is a technical document that describes a standard installation for a particular avionics system. That means aircraft manufacturers or airlines only have to design one installation, regardless of whose equipment they install. Another advantage for the airlines is that if a piece of equipment breaks down, and they do not have a spare, they can borrow equipment from another airline. Provided the equipment meets the ARINC Characteristic, it does not matter who made it.

The existence of an ARINC Characteristic for a new type of avionics gives it considerable credibility within the airline community. For several years we had been trying to persuade the Airlines Electronic Engineering Committee to authorize the writing of a characteristic for an airline Omega Navigation System. The airlines had been reluctant to do that until they became convinced the Omega System was ready.

Nevertheless, at each of the committee's annual general sessions, we tried to get an airline to propose a motion to begin work on an Omega ARINC Characteristic. Three times we tried, and three times we failed. However, we did not give up that easily.

By the autumn of 1974, Pan Am, TWA, and KLM were ready to sponsor an Omega ARINC Characteristic. We were confident that American Airlines, United Airlines, Air Canada, Air France, and others would also vote yes. It was enough to carry the motion.

We needed more than an ARINC Characteristic to help us sell Omega Navigation Systems to the airlines. An ARINC Characteristic does not deal with how well the system should perform, only with its physical installation. Performance is the domain of the world's aviation regulatory agencies. We knew it would not be long before we would have to participate in a committee set up by the Radio Technical Commission for Aeronautics.

The commission is an association of aeronautical organizations from both government and industry. As a U.S. organization, it deals mostly with topics of particular interest to the U.S., but its membership is international. Through various committees it tries to find mutually agreeable solutions to a wide range of technical problems facing the aeronautical community. It publishes reports and unofficial specifications. Agencies, such as the FAA, use or adapt those specifications in their own regulations.

Just two months after the airlines voted to begin work on an Omega ARINC Characteristic, the Radio Technical Commission for Aeronautics set up its own Omega committee. Its task was to write a specification that described the minimum performance and environmental standards for airborne Omega equipment. Many avionics manufacturers were on the committee, as well as airlines, aircraft manufacturers, the FAA, and other regulatory agencies. The first meeting took place in January 1975. After that, Peter Gasser and I made the trek to Washington, DC, every month to help with the committee's work. We were meeting the same people so often, competitors and customers alike, that a bonding started to occur. We solved many of our knotty technical problems, after our formal committee meetings, over drinks in the Greenery, a popular Washington bar.

The Omega ARINC Characteristic subcommittee started its work in May 1975. Consequently, for most of the next 12 months we had two industry committees to support. Of course, we had to continue with our marketing to the airlines, the military, and the regulatory agencies. We were too close to reaching our primary goal—the creation of a worldwide Omega market.

One day Peter would be on the bench helping Jean-Claude find 3 decibels of noise in our receiver, and the next day he would be in Washington, DC, debating whether we should design our equipment to operate at −30 or −40 degrees. I might come home from Rio de Janeiro via Los Angeles so that I could attend an ARINC meeting. Once there, I had to temporarily forget about selling to Varig, while we discussed issues such as whether the positive voltage of the signal to the left-right steering indicator should come from pin 67 or 44 of the bottom connector.

The ARINC Omega subcommittee document would become known as the ARINC-580 Omega Characteristic. That characteristic defined the installation requirements for an Omega Navigation System to work with a Doppler Navigation System.

We had estimated that about 500 aircraft worldwide would require a navigation system to replace the LORAN-A equipment. Some airlines, we knew, would select the Inertial Navigation Systems. Considering the fierce competitive situation, we thought that we would be doing well if we captured 30 percent of that market. That would be about 200 ship sets. We exceeded our projections by 100 percent, selling more than 400 for the LORAN-A replacement market alone.

During the next 10 years we sold 3000 sets of other versions of the product. We had to compete and work hard for each sale. Most sales were for 2 to 10 systems; only rarely did we have a customer who wanted 50 or more.

One title I considered for this chapter was "The Politics of Specification Writing." On the last Friday of each month Kieth Glegg, who was still our vice president, held a managers meeting. It was during one of those Friday sessions that he subjected me to one of his notorious caustic public remonstrations and lectures.

I reported on the first meeting of the Radio Technical Commission for Aeronautics and the strategy that we had adopted. I was in the middle of telling

Kieth that we intended to argue for a specification that we could easily meet. He stopped me midsentence; everyone cowered. Some appeared pleased that it was me, not them, who was about to get the monthly lecture; others felt sympathy for what was to come. I just sat in embarrassment. I could guess what was coming, and it was not going to be pleasant.

I do not remember Kieth's exact words—unpleasant situations are best forgotten—but I do remember the lesson. It went like this: "Graham, you are obviously too young to understand the politics of spec writing. Why on earth would you try to get the committee to agree to a specification against which any fool can design a product. If you have a spec that you can drive a city bus through, we will have so many competitors we may as well get out of the business. Do not pursue the strategy that you were about to suggest. I have been in this business since you were in diapers...." He went on, in this manner, for several minutes. By the end, I just wanted to be out of the room.

As usual, with Kieth, we received sound advice. He told us that he expected a specification against which only the best companies and the brightest engineers could design equipment. He suggested environmental and other parameters that were logical from the customers' perspective but that also would challenge our engineers. That was one of the very rare public lashings that I received from Kieth. Fortunately, once he had made his point, he did not hold such errors in judgment against an employee.

At the next meeting in Washington, we presented our suggestions. Several manufacturers objected, but eventually our proposals prevailed. One year and many trips to Washington later, we had a minimum performance specification for airborne Omega equipment. The FAA and Transport Canada accepted the recommendations of the Radio Technical Commission for Aeronautics. The United Kingdom's Civil Aviation Authority, usually wanting a made-in-England specification also adopted, with minor changes, the commission's recommendations. Soon the requirements of the specification were used throughout the world.

The deliberations of the Airlines Electronic Engineering Committee's ARINC-580 subcommittee also took 12 months. Again, we had to be careful to suggest a system's size and features that, while appropriate for the customer, were a technical challenge for the avionics manufacturers.

When the ARINC-580 subcommittee first met in May 1975, our work on the CMA-734 Omega System, our small second-generation system, was well under way. The control and display unit and the antenna unit were the same size as those units of our previous system, the CMA-719. Those two features of the systems did not change radically. However, the receiver-processor unit was half the width and shorter than that for our first system.

We went to the first meeting of the ARINC Omega subcommittee with the idea of persuading the airlines to specify a similarly small-sized receiver-processor unit. The airlines were anxious to have small lightweight systems, and so we thought we would be successful. Also, at that time, the airlines only wanted an Omega System to update, periodically, the navigation calculations of the Doppler Navigation System. They did not need other features, such as a steering signal to the aircraft's autopilot system or other signals to the pilots' instruments. The airlines did not want to pay for features that were not essential.

What we had not expected was that the other avionics manufacturers were not prepared to try to make such a small system. The airlines did not want the Canadian Marconi Company to be the only supplier of airborne Omega Navigation Systems, and so, as is typical with ARINC Characteristics, the committee selected a size acceptable to most manufacturers. It was 50 percent wider than the one we had proposed and little longer.

Back home our engineers looked for ways to use the CMA-734 without a major design change. General aviation customers, our principal target market for the CMA-734, do not usually worry about ARINC Characteristics. They prefer small and lightweight avionics rather than standard installations. We did not want to abandon the CMA-734 and a potentially lucrative market.

Peter Gasser came up with the solution. As usual, for Peter, it was simple and efficient. He decided that we should not alter the electronics cards or the way they were installed in the unit. Instead, our engineers simply put the CMA-734 circuit cards into a larger unit. We now had a lot of air space between the cards. That was an advantage because it meant the circuits remained cooler. Because the airline unit was also longer, we had an empty compartment in the front.

Having so much spare space was a good marketing feature. We could tell our prospective airline customers that our system could easily accommodate any additional circuits that they might want later. During the following years we were thankful for the extra space. One of the first circuits to go into that space was a receiver that could process the Navy's communication signals, which we used to backup the Omega System signals.

The ARINC Omega subcommittee had caused us to create a third Omega Navigation System. We called this airline version the CMA-740. We now had two design activities going on in our Omega Lab.

As the ARINC Omega subcommittee work progressed, we realized that some airlines might want to connect their Omega Systems to the aircraft autopilot and flight instruments. The Bendix Corporation agreed with our assessment, and so we suggested to the ARINC Omega subcommittee that we add those features to the characteristic that we were writing. The airlines with the strongest voice were those that wanted a minimum cost—therefore, no frills—system. They did not agree to our proposal. Nonetheless, Bendix, Tracor, Dynell, and Canadian Marconi decided to design such features and offer them as an option. We had no difficulty accommodating those features in our CMA-740 with all of its excess space.

Finally, in May 1976, the Airlines Electronic Engineering Committee approved the subcommittee's work and published the ARINC-580 Omega characteristic.

Teaming a Product and a Global Market

On Nov. 5, 1974, I walked into a Sheraton hotel in Washington, DC, to attend the Second Omega Symposium sponsored by the U.S. Institute of Navigation. Before I had a chance to register for the symposium, Fred Lackner and Irving Lublin stopped me. Fred was the Omega marketing manager for the Northrop Corporation, and Irving was the Omega project manager for the U.S. Naval Air Systems Command. I had met both of them many times.

Fred told me that he and Irving thought that the Omega community needed an association to encourage the exchange of information about the system. He asked if I would help them form such an association. I agreed readily. Initially, I was interested for the wrong reason. Mostly, I saw it as an opportunity to work more closely with Irving, which, I thought, should help us in our marketing efforts with the U.S. Navy. Also, I felt we could not afford to let others stake out an exclusive leadership position in an association dedicated to the Omega System.

Early in January, seven of us met in Washington, DC, to discuss the formation of an Omega association. In addition to me, Irving, and Fred, there were Cliff Barker and Hal Gershanoff, who worked for manufacturers of marine Omega receivers, Hank Walcott from the Bendix Corporation, and Joe Kasper from the Analytical Sciences Corporation, a technical consulting company that was carrying out Omega analyses. I was the only non-U.S. member of the group.

We were a mixed group, some of us competitors, some customers, but all brought together by a common cause. On many occasions members of an industry as a whole have to set aside, temporarily, competitive considerations to help each other. We decided that, whatever association we formed, it should not be a political lobbying group. We wanted a technically oriented association that would give military and civilian users of the Omega System a forum to discuss their requirements and experiences for everyone's benefit. We agreed that one of our first tasks should be to promote the Omega System as an international civilian system, because many civilian operators of aircraft and ships still viewed it as a U.S. military system.

Between January and April, we, the original seven, plus a few more people whom we had commandeered, met regularly to work out the details of forming the association. We were the founders. We had many debates over simple matters, such as name of the association and the design for its logo. Finally, we settled on registering it as the International Omega Association, and we agreed on a logo. Next came the more difficult tasks of writing a constitution, bylaws, and articles of incorporation and setting a dues structure and a plan for the next two years.

On April 18, 1975, just six months after Irving and Fred first approached me, the International Omega Association, a nonprofit organization, was open for business with a worldwide membership. Most of us who had helped to set up the association were on the Board of Directors. We were a working board. We ran the organization and organized its annual technical symposia and other activities. Irving Lublin was the association's first president.

The first annual meeting of the association was held in July 1976. The association has held major technical meetings, in various parts of the world, every year since. The association has made a major contribution to the international use and understanding of the Omega System. It also became a valuable marketing tool for Canadian Marconi and the other equipment manufacturers.

For most activities one can measure the rewards by the amount of effort he or she contributes. From the beginning, Canadian Marconi has remained active in the management of the association. I served continuously, in a variety of capacities, including president, on its Board of Directors until 1988. Henry Schlachta of Canadian Marconi has also served on the Board of Directors for many years.

When I agreed to help form the association, I did not appreciate the prestige or credibility that it would bring to Canadian Marconi. During the years when we were launching our Omega product line, we were seen as one of the leaders in this important international forum. I remain convinced that we secured many, especially foreign, airline contracts not only on the technical merits of our product but also on the credibility of the company in the Omega market. Our involvement and investment in the International Omega Association helped our image in many ways.

Once we committed to designing an airline standard Omega Navigation System, I was anxious to secure our position with an airframe manufacturer, and so I made regular trips to Boeing in Seattle. Late in 1975 Boeing agreed to offer our system as an option on its B-737 and later they offered the system as an option on the B-727 aircraft. It was over a year before we received our first order from Boeing, but during our negotiations Boeing introduced us to another industry standard, a document that would govern our business and our profitability in the airline market.

As the airline industry grew during the 1950s, so did the airlines' expectations of their suppliers. Airlines try to be self-sufficient, but to ensure that their suppliers continue to meet their obligations, the airlines produced a set of contractual guidelines they called *The World Airline Suppliers Guide*.

All of the world's airlines use that guide or versions of it that they have produced for their own circumstances. Boeing had to conform to the conditions of the guide and expected us, as its supplier, to also agree to those conditions. We were familiar with the guide, but when faced with having to make a binding contractual commitment, we were cautious. We did not want to promise to do something only to find out later that we could not honor that promise.

We were apprehensive, and so our negotiations with Boeing took several months. Our lack of knowledge about the airline industry was all too obvious. We need not have worried. During the next few years, we learned that our new customers, the airlines, are reasonable and practical. As we became established and more comfortable with our abilities in that market, we offered contractual commitments that far exceeded those we agreed to with Boeing.

Although airlines try to be self-sufficient, they expect their suppliers to provide prompt support when called upon. Aircraft that cannot fly because a piece of equipment has failed do not make money. Product support sets *The World Airline Suppliers Guide* apart from most other agreements for the sale of goods.

We had to guarantee the reliability of the equipment that we were providing. Manufacturers usually state reliability as an average time between failures. Typically, avionics manufacturers design their equipment so that it does not fail more than once every 3000 hours—the average annual flight hours for an airliner. A guarantee is worthless to a customer unless it is associated with a liability clause. Because airlines want to keep their aircraft flying, a financial liability is of little value. Instead, they require their suppliers to lend them, free of charge, additional equipment if what they bought does not meet the reliability guarantee. The airlines use a formula, which is written into the contract, to determine the number of systems that the supplier has to loan.

The airlines cannot afford to waste time removing and testing equipment they think has failed, only to find out that it is working correctly. Along with the failure guarantee, we had to commit to an unnecessary removal guarantee. That and similar guarantees influenced our equipment design. We also offered a three-year warranty, during which we repaired the equipment free of charge.

Through the provisions of *The World Airline Suppliers Guide*, we had to agree to provide training courses for the airline's flight crew and maintenance technicians. We had to agree to send field engineers to any part of the world if the airline was having a serious problem with our equipment. We had to agree to repair the equipment within 10 days or within 3 days in an emergency. We had to agree to respond within 24 hours if an airline could not fly its airplane because our equipment had failed and no spares were available.

Our airline customers expected us to provide those and other services around the clock, every day of the year. Also, we had to agree to provide that support for as long as the airline was using the equipment, which might be 20 years. Most of that support we had to provide at no cost other than the equipment purchase price. It is no wonder that when we first sat down to negotiate a contract with Boeing, we were nervous. In the years following our negotiations with Boeing, we learned how to satisfy those commitments without going bankrupt.

Soon after we had finished work on the ARINC-580 Omega characteristic, in May 1976, we started meeting more frequently with the FAA. It was writing the guidelines that the U.S. airlines would have to use to certify Omega equipment for use on the North Atlantic. The administration publishes those certification guidelines in a document called an Advisory Circular. In June 1977, it published the first of many advisory circulars governing the flight certification of Omega Navigation Systems.

The Canadian government, through DIPP, had provided us with some financial help to design our Omega Systems. The Canadian Armed Forces had given us the incentive to continue in the Omega business through its purchase of the CMA-719.

The United States, first through the Navy and then through the Coast Guard, along with six other nations, was providing the Omega ground stations. The International Civil Aviation Authority had unwittingly created an increased market interest in Omega by announcing that more stringent navigation standards would come into effect on the North Atlantic on Jan. 1, 1978. The FAA had helped our cause further by announcing the shutdown of the LORAN-A navigation system. We had worked with the Radio Technical Commission for Aeronautics and the Airlines Electronic Engineering Committee to develop specifications for airborne Omega equipment. We had also started to understand the contractual requirements of the airlines.

Almost eight years had passed since we had started in the Omega business. The military, general aviation, and commercial airlines were all poised to use this new navigation system. Back in the Canadian Marconi Omega Lab, our engineers were designing our second-generation Omega System, the CMA-734, for the general aviation market, and its derivative, the CMA-740, for the commercial airlines.

Industry and government had worked closely together to bring the market to this stage. It was now up to us to capitalize on this opportunity.

10

Pan American World Airways, Thank You!

Leaders don't <u>force</u> other people to go along with them. They <u>bring</u> them along. Leaders get commitment from others by giving it themselves, by building an environment that encourages creativity, and by operating with honesty and fairness.
—United Technologies Corporation
(from a guest editorial)

They are all gone now, friends and business colleagues, in early retirement or working for other companies.

On Dec. 4, 1991, the grand old lady of civil aviation, Pan Am, ceased to exist. The airline had been a pioneer of commercial aviation and for many years the flag carrier of the United States. Anyone who knew Pan Am during its 64 years of existence must have been greatly saddened when the airline succumbed to almost a decade of fierce competition and financial woes. I am sure that many employees, equipment suppliers, and frequent travelers shed more than one tear on Dec. 4. It was like losing a friend who had suffered from terminal cancer.

The *Washington Post* wrote, "Pan American World Airways went out of business yesterday, swept into history after 64 years, by wrenching changes in the industry it helped to create."

In 1927 Pan Am introduced the world to the Clipper Ships, and the age of commercial aviation was born. It became the U.S. flag carrier and developed a route structure that transported people, mail, and freight to all corners of the world.

By 1975, when we first started marketing our Omega Navigation System to Pan Am, it was revered, and rightfully so, around the world. It had been and continued to be a pioneer. As *Aviation Week and Space Technology* wrote in a

Dec. 16, 1991, article: "By that time [mid 1970s], most of the world's airlines were reaping the benefits of Pan Am's pioneering work." It had been the launch customer for the world's second commercial jet aircraft, Boeing's B-707, and for the world's first jumbo jet, the B-747.

Pan Am took its place in the industry seriously. It helped many manufacturers develop products for its own and other airlines' use. Sadly, today few if any companies are prepared to make that type of investment on behalf of an entire industry. Pan Am also pioneered commercial airline use of Omega Navigation Systems.

It was on the world's air routes, not on the domestic ones, where Pan Am conducted its business. It did not matter if you spoke Chinese, Japanese, Portuguese, Spanish, German, Russian, or one of more than 50 other languages. If you traveled, you knew the name Pan American World Airways. Everyone associated with Pan Am benefited from that association. We, at Canadian Marconi, wanted to be associated with Pan Am, too.

We wanted to break into the commercial airline market. I remained particularly committed to that market. I felt equally strongly that it was not healthy for our company to rely on military or other similar government contracts. Others at Canadian Marconi, particularly Peter Gasser, felt as I did.

Pan Am was one of the first airlines to react to the news, late in 1974, that new navigation standards would come into effect on the North Atlantic in January 1978. It had 50 B-707 aircraft operating on the North Atlantic and other world routes that were equipped with Doppler and LORAN-A navigation systems. Because the FAA intended to close down the LORAN-A ground transmitters, Pan Am only had two choices. It could buy Inertial Navigation Systems, the same systems installed on its new B-747 aircraft, or alternatively it could buy Omega Navigation Systems, which were then about one-quarter of the cost of the Inertial Navigation Systems, and use them with the Doppler systems.

Pan Am delayed its decision because there was still much that it wanted to know about the performance of the Omega System. It only had three years to buy, install, and certify the new navigation systems for its B-707 aircraft. It knew that it would take nine months to buy the equipment and one year to install. And so Pan Am had about one year before it had to decide which equipment to use. Omega Navigation Systems were attractive because of their low cost and because Pan Am expected to retire their B-707 fleet within 10 years. KLM and TWA were in the same position as Pan Am. Our target market was clear, and so we started to make regular trips to all three airlines.

The four years between 1975 and 1979 were the busiest ones for me and Christina, though they were years of excitement, challenge, and success. Both of our children were born during that time, and I became the man who showed up periodically with toys from all over the world—the absentee father's guilt presents.

During 1975 my colleagues at Canadian Marconi and in the Omega community and I worked on the Omega subcommittees of the Airlines Electronic Engineering Committee and the Radio Technical Commission for Aeronautics. Both committees met every month for three days each. We also made regular visits to Pan Am in New York, TWA in Kansas City and New York, and KLM in Amsterdam. I visited Boeing on numerous occasions. I returned to Varig and Japan Airlines. We also started to visit other airline companies, such as Air France, American Airlines, Eastern Airlines, and British Airways. We did not expect those companies to be early customers for our Omega Systems, but we knew that we had to start our marketing early. Only with sustained marketing could we establish the company's credentials in this market, which is conservative and wary of new and unknown suppliers, especially suppliers like Canadian Marconi, which was both unknown and had a generically new product.

Early in 1975, Don Bowyer, of Marconi Avionics UK, and I visited Moscow. We went at the invitation of the Soviet airline, Aeroflot, because it wanted Omega Navigation Systems for its Ilyushin IL-62 aircraft flying the North Atlantic. Export regulations prevented us from offering the new airline system we were designing, and so we could only talk about supplying our first generation system.

It was a fascinating week; Don and I looked for electronic eavesdropping devices in our hotel rooms and were ever alert for suspicious looking people. We were suffering from an acute dose of paranoia. We had no reason for that paranoia, except for the security briefings that we had received before our departure for Moscow.

Our Soviet hosts were hospitable and friendly and gave us the impression that we would soon be doing business with each other. We drank too much vodka and had difficulty finding anything to eat. Provided we stayed within the city limits, we enjoyed more freedom than we expected. Still, our paranoia caused us to think that we were always being followed.

Our Soviet hosts placed a car and chauffeur at our disposal, but frequently Don and I would just walk the streets and ride the subway. During one evening excursion, we became lost and walked for about three hours. Finally, we decided to catch a taxi back to our hotel. We hailed the first taxi we saw—our chauffeur was driving!

Our marketing was becoming more focused. Despite our visits to several airlines and airframe manufacturers, we concentrated on Pan Am, KLM, and TWA. By the middle of 1975 Pan Am was ready to try an Omega System in an aircraft. We readily agreed to loan it a system, though the only equipment that we had available was the first-generation CMA-719. The CMA-734 was still in the prototype stage, and its derivative for the airlines, the CMA-740, was still on the drawing board.

The CMA-719 was adequate for Pan Am's purposes. First it wanted to get some experience with the Omega System. Later, it would test various products.

Dynell, which within a year the U.S. Air Force would select as its Omega Systems supplier, also loaned equipment to Pan Am.

To navigate, our system needed to receive only one signal from three stations, but to do that, it needed a complex software program. Dynell had found a way to reduce the complexity of its product by mixing two signals from each of three stations. Though that approach produced a lower-cost system, it meant that it had to receive two signals, not just one, from each station. We believed that this was a serious disadvantage because we could not be assured of receiving two signals from a station.

When an Omega System does not receive enough signals to calculate its position, it automatically goes into a mode called dead reckoning and uses the aircraft's airspeed and compass information to navigate. The equipment alerts the pilot to that less accurate mode of navigation by a light on the control and display unit.

We expected the Dynell system to go into its dead reckoning mode frequently during the Pan Am tests. The airline, however, flew the equipment over the North Atlantic where the Omega signal coverage was good, and so both systems performed well. The tests were good enough to convince Pan Am that it should seriously consider Omega Navigation Systems for its B-707 fleet.

We had continued to visit the FAA, and in the summer of 1975 it decided to check the technique that Dynell had adopted for its system. The technique was so novel that the FAA wanted to check its appropriateness itself. To ensure an unprejudiced test, it intended to buy a system and not accept one on loan. It awarded contracts to a systems engineering company and Pan Am to help with the test.

The FAA issued a competitive Request for a Proposal for the Omega equipment. We decided that this was an opportunity to prove the superiority of our approach to Omega signal processing. We offered a system that would work like the Dynell system but that with the flip of a switch would operate more conventionally, like our own design.

The FAA awarded Canadian Marconi the contract, and we modified one of our few prototype CMA-740 systems. The flight tests, which Pan Am carried out on both its North Atlantic and Pacific routes, lasted from August 1975 to May 1976.

When we decided to bid aggressively for the FAA's test program, we had more design work under way than we could handle. Several of our engineers bemoaned the marketing department, me, for once again trying to bring in work that would detract from their main activity. However, we could see many benefits to participating in this demonstration program. Principally, it would give us additional exposure to Pan Am. Of course, if the system did not work well, then we would have a real problem. It would be difficult for us to regain our credibility. However, this was not a time to be timid.

Fortunately, the system worked, at least with the signal processing approach that we favored. That test convinced Pan Am to fit Omega Navigation Systems into its B-707 aircraft. All that remained was for it to select a supplier.

By that time, we had completed the ARINC-580 Omega Characteristic, the specification that defined the size, installation, and basic functions of an airline standard Omega Navigation System. The airlines that intended to fit Omega equipment to their aircraft could now start preparing the installations, even though they had not selected a supplier.

The cooperative project between Pan Am and the FAA, and our involvement, gave us many opportunities to visit both organizations. Regular visits to prospective customers are a key ingredient of any marketing plan. The problem is to find good reasons to visit often enough, without wasting their time with meaningless visits. We had an ideal situation.

Those trips to Pan Am gave us an opportunity to get to know several Pan Am employees, people who would be influential in the selection of an Omega System supplier. Peter Gasser, by then our senior project engineer, and Jacques Raia, Pan Am's avionics engineering supervisor, developed a close working relationship born out of mutual respect for one another's other's technical integrity. Most important, for us, Jacques trusted Peter. Jacques knew that if Pan Am selected the Canadian Marconi Omega System, he could rely on Peter to do everything in his power to satisfy the airline. Would that be enough, though, to sway Pan Am?

Initially, our relationships with Pan Am employees were friendly but professional. Real friendships came later, after Pan American selected its Omega equipment supplier.

Our friends and mentors at Pan Am were Jacques Raia, Bill King, manager of engineering, and Pat Reynolds, the last of Pan Am's navigators. Those people taught us the most about doing business in the airline industry. Typical of this commercial aviation pioneer, its employees accepted the role of teacher.

Because Pat was a navigator, during the flight tests and when Pan Am introduced the Omega equipment into service, he sat in cockpit jump seats, flying all over the world, assessing the performance of the equipment. Almost single-handedly he managed all of the tasks that an airline must perform to bring a new navigation system into service. Most people would have found that job exhausting, but Pat, who remained at Pan Am well past the normal retirement age, exhibited limitless dedication and energy.

Pan Am appeared willing to give us an opportunity to prove ourselves. At least our continued involvement would help keep the competition going. Our principal competitor was the large Bendix Corporation. It had been dealing with Pan Am for many years and was a trusted U.S. supplier. The competition also included Dynell Electronics and the Tracor Corporation, both U.S. companies. Our first competitor, the Northrop Corporation, had withdrawn from further competition in the Omega business. Litton's Aero Products Division soon joined in the competition. Global Navigation Systems Inc. and Communications Components Corporation were selling navigation systems that used the U.S. Navy's communication signals. Also, two U.S. companies, Litton and Delco, were selling Inertial Navigation Systems. All of those companies were competing with us for the same market.

Canadian Marconi, though located in nearby friendly Canada, was a lone foreign company competing against U.S. corporations. We were also a small company by U.S. standards. To make matters worse, our only previous involvement in the airline industry had been a short one 15 years earlier ending when the company had withdrawn to concentrate on the military market.

KLM remembered our first involvement with the airlines. It still operated Canadian Marconi Doppler Navigation Systems on its Douglas DC-8 aircraft. However, it never forgave the company for withdrawing from the airline market. KLM proved to be the most formidable prospective customer that I ever had to face.

We started to pursue KLM as soon as work on the ARINC-580 Omega Characteristic started in May 1975. Trips to Amsterdam became a frequent occurrence. During the next 24 months I did everything I could think of to persuade KLM that this time Canadian Marconi had a long-term commitment to the airline market.

John Mohr, KLM's avionics engineering supervisor, had been a junior engineer when Marconi sold Doppler Navigation Systems to the airlines. Eventually, I was sure I had convinced him that this time we would not withdraw from the market.

KLM's biggest complaint was that we had not completed all the maintenance manuals for the Doppler system. Despite this, the airline had long ago become self-sufficient in the maintenance of the Doppler system. It even admitted that the system worked well.

Although John Mohr may have been willing to give us another chance, others at KLM were more reluctant. When we finally provided them with a proposal for the Omega Systems, we included a clause that stated we would provide them with support engineers, at no cost, until we had delivered all the manuals. We also agreed to prepare the Doppler manuals that they still wanted.

Peter Gasser and I made so many trips to TWA in Kansas City that we considered starting our own video conferencing company. In 1975 we realized that the technology for such an idea was available. We envisaged putting a video conferencing center in each city. Then if two parties wanted to have a face-to-face meeting, all they would have to do was book time at their center. Today such a service is available; in 1975 it was an idea that we only talked about on the airplanes to and from Kansas City.

TWA was the other major airline that had to replace the LORAN-A equipment by Jan. 1, 1978. We felt that we were making good progress with TWA. Its only knowledge of us, though, was through those visits and contacts with several of its engineers at meetings of the Airlines Electronic Engineering Committee.

By the summer of 1976 all three airlines, Pan Am, TWA and KLM, were carrying out flight tests of the competing Omega equipment from Bendix, Dynell, Tracor, and Canadian Marconi. We often faced a situation where a military or commercial operator intended to flight test all the competing products one after the other. Then we would have debates about whether it was better to be first, in the middle, or last in the test sequence. It does not matter because customers make decisions based on what they want, what they can afford, how the product performs, and whose proposal most closely meets their requirements.

Pan Am began flying our CMA-740 Omega System in one of its B-707 aircraft in July 1976. That flight test program, our third with Pan Am, was the one that really mattered. After that test Pan Am would compare the results from our system with those they had collected from our competitors' equipment.

We had a harrowing meeting with Pan Am shortly before the start of those flight tests. Bill King and Jacques Raia knew we only had a prototype system, but they were confident we could complete the design and put it into production. However, we needed to convince their director of engineering.

After we showed him our prototype equipment, he gave Jacques and Bill a long lecture about Pan Am being a commercial airline, not a test center for prototype avionics. Our market nearly came crashing down around us during that one meeting. There were several tense moments. Jacques and Peter put forward a convincing argument, and finally the director reluctantly agreed to let the flight test continue. We had come far too close to a disaster.

Pan Am flew our system for nine months, between July 1976 and March 1977. In 1976 KLM flight tested another CMA-740 system that our engineers had hurriedly put together, and TWA tested a third system. We only had two spare systems, and so whenever one failed, our engineers worked continuously until they fixed the problem. Sometimes we would have an unexplained navigation error, or the system would temporarily stop receiving Omega signals. Then our field engineers, or Pierre Fournier, from our training department, or one of the Omega engineers would fly with the aircraft to find the cause of the problem. We worked together as a team with just one goal—a launch customer.

In July 1976 the Omega station in Argentina came on the air. Finally, we now had seven of the eight Omega stations on the air. The eighth station, in Australia, came on the air in 1982, some 10 years later than the U.S. Navy had originally planned. After waiting eight years for the Omega stations, we celebrated the commissioning of the station in Argentina with some considerable relief. Even the station in Liberia, which had replaced the experimental station in Trinidad, had been operating since February.

The flight tests of our CMA-740 Omega System by Pan Am, TWA, and KLM were successful. The system's accuracy on the North Atlantic was usually better than 2 nautical miles and most often within 1.5 nautical miles.

As I remember, we only had one complete failure of the equipment during those flight tests and that was during a Pan Am flight. The equipment failed on the way to Europe, and we got the message while the aircraft was still airborne.

The aircraft would only be on the ground in Europe for 12 hours, and so we decided to use that opportunity to prove we could respond to an emergency. We quickly prepared a replacement set of equipment and the required customs paperwork. Within a few hours, one of our field engineers was on a commercial flight to Europe with a replacement CMA-740 Omega System as his only baggage. He arrived just in time to exchange the equipment on the Pan Am aircraft and then sat in the cockpit jump seat for the return trip. We proved our point. Several months later Pan Am employees told us that they had remembered our quick response when they had selected their Omega supplier.

We frequently used cockpit jump seats during those flight tests so that we could witness the equipment in operation or help track down the cause of a problem. Usually, it was our field engineers or instructors who spent long hours in the cockpit, but we also sent our design engineers and our marketing personnel. Employees have a stronger sense of involvement and commitment to their work if they can see their products working in the customer's environment. Occasionally, we could mix a little pleasure with that type of trip.

Several years after my father retired from Marconi in England, my parents decided to emigrate to New Zealand, where my sister was living. They left England in 1976 at the time TWA was flight testing our system on the North Atlantic. The airline graciously offered to fly me over, in the jump seat, so that I could say goodbye to my parents. The trip also gave me an opportunity to see, for the first time, the CMA-740 in operation over the North Atlantic. After my parents settled in New Zealand, I naturally became interested in securing business in that part of the world. When I became marketing manager, I tried to assign my staff to countries and regions where they also had a personal incentive to succeed. We secured business throughout the Pacific, including New Zealand and nearby Fiji. We might not have been so successful if my parents had not emigrated to New Zealand.

In addition to the flight tests by the three airlines, which would be the first commercial users of Omega equipment, 1976 was a busy year in other respects. We continued to work with the Radio Technical Commission for Aeronautics, the Airlines Electronic Engineering Committee, and the FAA, developing various specifications and flight certification criteria for Omega Systems. The engineers in the Omega Lab, completed most of the work on the new CMA-734 Omega System, having flight tested the equipment in the company aircraft during the previous year. They released the product for production. Now the engineers had to solve all the small manufacturing problems that came up daily.

With the CMA-734 in production, we could boast that we had the world's smallest Omega System, a fact that Carl Hanes and Tom Kelly emphasized as they continued to pursue the general aviation market. They were having some modest successes. They also reached agreements with several general aviation fixed base operators. We had about 20 general aviation customers anxiously waiting for the first CMA-734 systems to come off the production line. Carl and Tom continued to succeed, and orders for one or two systems started to come in regularly. We were now committed to that market.

Our engineers had to cope with more than just completing the CMA-734 and nursing it into production. They were also completing the design of the CMA-740 airline Omega System, which was a repackaged CMA-734 with a few additional features. A year earlier they had started to design a receiver that could process the U.S. Navy's very low frequency communication signals. The CMA-740 processor would use those signals when there were insufficient Omega signals for navigation.

At the first annual meeting of the International Omega Association, which we held in Washington, DC, I presented a paper titled "An Automatic Omega Navigation System with V.L.F. (U.S.N.) Augmentation." The paper described the CMA-740 that we were developing for the airlines and how we were incorporating the special receiver into the system to provide backup navigation.

My presentation was the first time that an Omega equipment supplier had publicly endorsed the use of the communication signals for navigation. Until very recently, we all had been trying to convince the FAA, and anyone else who would listen to us, to prohibit the use of those signals for navigation. The FAA had compromised by telling the airplane operators they could only use the communication signals for backup navigation. We had accepted the compromise, in part because we were still trying to convince the U.S. Air Force that our equipment, with that dual receiver capability, was better than the others it was considering.

Early in 1977, during our negotiations with Pan Am, Bill King and Jacques Raia visited Canadian Marconi. They wanted to meet the company's senior management and to see our engineering, production, training, and product support facilities. The inspection was one of several criteria they would consider when they selected their Omega System supplier. It was a critical visit.

I believe that the visit and their discussions with John Simons, our vice president, and Ervin Spinner, now our marketing manager, were a turning point for Pan Am. Up to this point, they had come to respect Peter Gasser and our engineering team. We had proved that our system worked. Also, we had shown them that we supported our product and that we could respond to emergencies. Many times they had witnessed the commitment of our entire Omega team. Bill King told me on several occasions that he was impressed that Canadian Marconi had supported for so long the Airlines Electronic Engineering Committee. We had considered our participation in its activities an essential part of doing business with the airlines. I think what impressed Bill most was that, although we were a small company compared with our competition, we contributed to the technical discussions as if we were a large firm with more resources.

It was not difficult to show Bill and Jacques that we had the facilities and staff to produce and to support our product, though, by most standards, we were understaffed. Colloquially, one could say that we were a "lean but not a mean operation."

After just a short time in John Simons' presence, his professionalism, integrity, and commitment to the airline market were evident to our visitors. It demonstrated the influence a company executive can have in the market. Similarly, Ervin Spinner's compassion and sense of decency quickly came across. If Pan Am chose Canadian Marconi, its representatives realized that Ervin would put a customer above every other consideration.

With John Simons and Ervin Spinner in the top jobs and Peter Gasser leading the engineering effort, Pan Am knew that the company would be a reliable supplier. John Simons had promoted Michel Galipeau from manager of the Omega Systems product to manager of our quality and product assurance activities. He had also promoted Lionel Leveille to head the division's Avionics Product Support Group. Two seasoned professionals held two critical jobs. Pan Am's representatives were confident having Michel, whom they knew and trusted, lead our quality assurance program. They would soon learn to appreciate Lionel's skills and commitment to product support.

During that year of frenetic activity with the airlines, we continued to visit military agencies. Our heritage was the military, and we wanted to maintain a strong position in that market. Despite our initial lack of success with the U.S. Navy, which was flying the Northrop system, we kept it informed of our new Omega product development. We knew that one day, perhaps several years away, it would want to replace the Northrop systems or to equip more aircraft with Omega Systems. Success goes to those who cover all eventualities well in advance.

The Lockheed Aircraft company continued to manufacture the world's military transport workhorse, the C-130, which is used by many military agencies throughout the world. We visited Lockheed on a regular basis to discuss the advantages of fitting C-130s with our Omega equipment.

We continued our marketing to Britain's Royal Air Force. In addition, we started discussions with Hawker Siddeley Aircraft in England and De Havilland of Canada. Both produced various military transport aircraft.

By December 1976 the three airlines—KLM, Pan Am, and TWA—that had to fit Omega Navigation Systems to their airplanes by Jan. 1, 1978, still had to select a supplier. Pan Am decided to fit each of its B-707 airplanes with two Omega Systems, even though the regulations required only one, working with a Doppler Navigation System. Pan Am was anticipating the day when the FAA approved the use of the Omega System as a primary navigation system. Then it would need two systems on each airplane. Pan Am's decision meant that it would need more than 100 sets of equipment in less than 12 months. KLM and TWA needed far fewer systems.

To meet such a delivery schedule, we knew we would have to start building CMA-740 equipment right away, without a contract. It was too late to buy the parts we would need, and so we had to consider converting most of the CMA-734 systems that were already in production.

Peter recalls calling Jacques Raia on Christmas Eve. By then they could talk quite candidly with each other, such was their mutual trust. He told Jacques that he had just diverted our production effort, at some considerable risk, just in case Pan Am should select the Canadian Marconi Omega. He wanted to know if he had made the right decision because he was laying awake at night. Pan Am had not made its decision, but we assumed its employees had held several meetings on the subject. Whatever the situation Jacques had to be circumspect. In answer to Peter's question, Jacques simply said, "Peter, sleep well and enjoy Christmas." It was the first sign that we had received that we were a strong contender.

The winter months of 1977 were a period of intense negotiations with the three airlines we had been courting for more than two years. They had issued their formal Request for Proposals. Those were the airlines' own variants of *The World Airline Suppliers Guide*, with which we were by now quite familiar.

I made many trips to Amsterdam because KLM was still unsure whether it could trust Canadian Marconi. At its request we wrote several unique clauses into the proposed contract to cover all its specific concerns.

We sensed that TWA was uncomfortable about buying a crucial navigation system from a foreign company, one with which it had no previous experience. Again, we offered special clauses that we hoped would allay those concerns.

Pan Am, true to its reputation, was the most helpful. By the time we submitted our proposal, we were confident that the competition was between Bendix and us. Bendix had sold several products to Pan Am in the past and enjoyed a good relationship with the airline. Despite Jacques' comment to Peter on Christmas Eve, we were not assured of the Pan Am contract. Later, we learned that the competition between us and Bendix had been excruciatingly close. It was only after several senior management meetings that Pan American made a decision, which was an unusual occurrence for such a small, by its standards, procurement activity. It was suggestive of how fierce the competition was for that first major Omega contract.

Ervin Spinner recalls that after one of our last and particularly intense negotiation sessions, Bill King put his arm around Ervin and said, "You know, Ervin, Canadian Marconi is really ignorant about the airline industry, but we are going to help you learn."

We were now down to the hard issues. Pan Am's primary concern with our proposal was that, although we had agreed to product reliability and performance guarantees, it still had to ship equipment to Canada for repair, at least until it was in a position to repair the equipment itself, which the airline did not want to do until after the three-year warranty period. At that session Pan Am asked us to consider setting up a repair center in New York. Pan Am even gave us some suggestions. We had two choices. We could hire a local company, or we could set up our own local company to provide that service.

Ervin Spinner and Lionel Leveille dropped everything they were doing to concentrate on researching the most effective approach. Finally, they decided to enter into an agreement with Butler Aviation, with whom we were already working in the general aviation market. The arrangement meant that we would have to make an additional investment, but it was cheaper than setting up our own repair facility. We would have to provide Butler Aviation with all the special Omega test equipment that it would require. We would have to train its technicians and provide them with spare parts and systems. Then we would have to pay them for their time repairing Pan Am equipment on our behalf.

It was obvious that if we were serious about the airline market, and Pan Am was our door to that market, we would have to make that investment. It was money well spent.

A few years later Ervin and Lionel once again analyzed our support situation. They concluded that the time had come to set up our own facility. Within a short time they had selected a facility, hired staff, and equipped them with all the equipment they required to repair the Omega equipment. From that small beginning evolved C.M.C. Electronics Incorporated, a U.S. subsidiary of Canadian Marconi.

Our market intelligence suggested that the bids that the three airlines had received from Bendix, Tracor, Dynell, and Canadian Marconi were very close. That did not surprise us because the procurement had been highly competitive. Dynell's bid concerned us. The U.S. Air Force had recently awarded it a contract. Its bid had been more than 10 percent lower than the next lowest competing bid. Such a large price difference was very surprising in that type of procurement.

Once we completed the negotiations, all we could do was wait—we were very nervous. It was more than eight years since we had started to design our first Omega System. The delays in the ground stations had eroded confidence in the system. As a result, we had been unable to achieve an acceptable return on our initial investment with our first-generation equipment. True, the Canadian forces had bought a large quantity, but that was not enough to justify the investment.

Canadian Marconi had invested in the design of two second-generation Omega products. After eight years of investment in the Omega market, our future would now be decided by one of three airlines. Although Pan Am would be, by far, the biggest prize, we could continue in the business if we had an order from any one of the three. Pan Am would open a large international market. TWA would predominantly open the North American market to its supplier, and KLM would help a company to sell in the European market. To remain in the market we had to win at least one of the three competitions, which the airlines were now debating behind closed doors in New York, Kansas City and Amsterdam.

We found it difficult to concentrate on anything for the first two weeks of April.

TWA was the first to call: "We would like to thank you for your cooperation, but we have selected Bendix." Our worst fears had been realized. Surely Bendix was going to capture the entire three-airline market.

During the third week of April, Pan Am called: "Congratulations. How quickly can you deliver 105 systems?" Although it was only 9 am, we celebrated with the champagne we had stockpiled—just in case.

One hour after the call from Pan Am, I received a transatlantic call from John Mohr at KLM: "We would like to thank you for your proposal, but we have selected Dynell." The news disappointed me, but we were deliriously happy about the Pan Am contract. I had a feeling of personal failure because I had conducted most of the marketing to KLM on my own. I replied, "John, I'm also really sorry because we really are serious about the airline market. I hope that we can do business with each other in the future." I could not help adding: "By the way, Pan Am just called to let us know that it has selected our system. It has ordered 105 systems." The KLM requirement was small by comparison.

We had secured the largest airline Omega contract, though we were a Canadian company, and our customer was the United States' national flag carrier.

11

"One Sure Customer Can Lead to Another"

Success is simply a matter of luck. Ask any failure.
—Earl Wilson

Pan Am, TWA, and KLM had selected three different suppliers: Canadian Marconi, Bendix, and Dynell. Only Litton and Tracor were without a launch customer. Litton had been a late entrant into the competitions and too late to have any serious influence. It had its eyes on American Airlines and other U.S. airlines that were operating its Inertial Navigation Systems. We did not know if Tracor would stay in the market; it did and remained a strong competitor.

The KLM decision distressed me for some time. I knew that we had been sincere in our proposal and the negotiations. We would have done everything possible to satisfy KLM if it had bought our product. We did not intend to repeat the errors of our past with KLM or any other customer.

Companies do not always get a second chance, and we were not getting it this time. Obviously, I had not been successful in conveying our message. I was convinced that Dynell would soon become consumed with its U.S. Air Force contract. I wondered if it would make the same error that Marconi had made many years earlier when it also secured a large military contract.

Eighteen months later, at the general session of the Airlines Electronic Engineering Committee, I saw a note on the message board. It read: "If a representative of Dynell is here, would he please contact John Mohr." Every company involved in airline avionics is meant to be at those annual meetings. Where was Dynell? A few years later the company was bought by a larger concern. About one year after the takeover, it withdrew from the Omega market, but only after it had beaten us in a few more competitions, particularly in Europe.

In the meantime, we had to make sure that Pan Am could not complain about the Canadian Marconi service. It never did, and soon it was holding us up as an example of how a company should perform.

Peter Gasser, our engineers, and the quality assurance and manufacturing departments worked wonders to deliver 105 Omega Systems to Pan Am. They completed and delivered all the systems between April and December 1977.

Although we had started building CMA-740 equipment in January, in anticipation of at least one airline order, it was not soon enough. To meet the Pan Am delivery schedule we had to divert electronic assemblies from the CMA-734 production line. That meant we had to do something we were desperately trying not to do. We delayed some general aviation customer shipments.

While we were manufacturing the CMA-740 for Pan Am, we had to subject some units to rigorous environmental and electronic tests. Transport Canada required those tests to certify that the equipment was suitable for installation in commercial aircraft. Pan Am also had to rush its installation work. It had already wired several aircraft to accept dual Omega Navigation System installations. Its technicians began to wire the remaining 100 plus B-707 airplanes in its fleet. As soon as equipment came off our assembly line, we shipped it to Pan Am.

Before Pan Am could remove the LORAN-A equipment and officially fly with the Omega Systems, it had to prove the system's performance to the FAA. By June 1977 the FAA had completed the preparation, with our help, of the first Omega North Atlantic Minimum Navigation Performance Standards Advisory Circular. That document gave airlines guidelines about how to certify their Omega installations on North Atlantic routes. Pan Am, which was still installing the Omega Systems into its B-707 fleet, immediately began to collect navigation performance data.

Pat Reynolds flew across the Atlantic 64 times collecting data. Often an FAA inspector sat with him in the cockpit. To test the system under bad signal conditions, they prevented the system from receiving some stations by inserting a special code through the pilot's control unit. Mostly, though, they compared the system's navigational accuracy against a ground radar on each side of the Atlantic and against other navigation systems.

Pan Am, unlike most other airlines at that time, flew throughout the world. Therefore, it needed an Omega–Doppler system combination that it could use on all of its B-707 routes. Although the new navigation standards did not apply in the South Atlantic or the Pacific, Pan Am collected data in those regions. It was a nerve-racking time for those of us on the Marconi Omega team. If the system did not meet the navigation standards required by the FAA, it would ground the Pan Am B-707 fleet.

Pat, an intrepid master navigator, had the most difficult task. At age 65, or thereabouts, Pat saw the certification of the Omega System in Pan Am's fleet as his last act in a very distinguished career. He personally participated in all the certification flights. He not only flew the North Atlantic but also made regular trips to South America and across the Pacific ocean.

Our collective efforts were rewarded in September 1977 when the FAA granted Pan Am approval for its Marconi CMA-740 Omega System and Bendix–Doppler system combination. On the North Atlantic the CMA-740 system had shown that it could navigate to within 1.5 nautical miles most of the time. That was within the accuracy required for the new North Atlantic navigation standards. The FAA also allowed Pan Am to use the system on its South Atlantic and Pacific routes.

Pan Am was the first airline to receive North Atlantic certification using Omega equipment. The South Atlantic and Pacific certification was an added bonus. It also helped our marketing effort. The Pan Am certification program was typical, and we had to support similar certification programs with all our customers. Six months later, in March 1978, Pan Am became the first airline authorized to use Omega, with Doppler, for navigation throughout the world.

By the autumn of 1978, the Federal Aviation Administration had enough confidence in the Omega System to issue procedures against which airlines could apply to use it as the only means of navigation. Pan Am was the second carrier, and the first major commercial carrier, to receive that certification. Ports of Call Travel Club was the first operator to receive certification and also use Canadian Marconi Omega equipment.

At a ceremony on the production floor at Canadian Marconi in December 1977, John Simons, our vice president, handed Bill King the 105th set of CMA-740 Omega navigation equipment—suitably wrapped with a blue ribbon. Bill King told the entire team: "Pan American has never regretted its decision to select Marconi. You have done a remarkable job in delivering all systems on schedule."

Jacques Raia of Pan Am and Peter Gasser looked on and winked at each other. Only they knew just how remarkable was the accomplishment. Both the Canadian Marconi and Pan Am teams had worked long hours to get to this point.

The marketing team did not sit back while the engineers and the production staff worked to complete the Pan Am deliveries and flight certification. While we were pursuing the three airlines leading the market to replace their LORAN-A equipment, we were also pursuing many other prospective customers. We never slowed down. We knew that during this period we had to secure a position in the civilian market.

Securing a launch customer had been crucial to our marketing effort. The Pan Am contract gave us the best possible launch. Not only was Pan Am well known throughout the world, but also its contract was large enough to give us a strong production base from which to start.

Boeing was now offering our CMA-740 Omega System as an option on its B-727 and B-737 airplanes. By the time of the Pan Am contract we had already received orders from Boeing, and it was expecting equipment in August—right in the middle of our deliveries to Pan Am.

Though our engineers were working frantically to complete the CMA-740 design and to help put it into production, there was other work for them. Our marketing efforts resulted in more requests for evaluations by prospective customers. Those evaluations always put an additional burden on our engineering staff. In addition, some airlines, realizing that the Omega System had greater potential than they first thought, wanted to specify a system with more capabilities than the one described in the ARINC-580 Omega Characteristic, against which we had designed our CMA-740 system.

The Airlines Electronic Engineering Committee authorized a subcommittee to define a second-generation airline Omega System. A second-generation system? We had not yet sold the first one. Despite that invitation to continue with our investment, when we could least afford it, we were encouraged by the situation. Finally, the world's airlines, not just the few that had to replace the LORAN-A equipment, were taking the Omega System seriously.

Consequently, in addition to all our other work, we were once again making monthly trips to various U.S. cities, and helping to write another airborne Omega System specification. It became known as the ARINC-599 Omega Characteristic. The subcommittee published it in October 1977, just seven months after our Pan Am contract. We must have done a good job because this specification has withstood the test of time. Twenty years later it is still the standard installation specification for an airborne Omega Navigation System, although, internally, the equipment has become more complex and has greater capabilities.

While half of the members of our Omega engineering team worked on the CMA-740 system for Pan Am and Boeing, the other half worked on the new system. Fortunately, the CMA-740 had spare space in its main electronics unit. Our next system, which we called the CMA-771, was a variant of the CMA-740, but with some important differences. The ARINC-580 Omega Characteristic had described a basic system with only a minimum of interfaces to other aircraft systems. Conversely, the ARINC-599 Characteristic called for many more outputs. With those additional signals the system could drive various pointers and digital displays in the cockpit. The new Omega equipment also provided electrical signals to the aircraft autopilot, the critical aircraft system that automatically flies an aircraft on a preset course. An important new feature was a special electronic digital interface that allowed two Omega Systems in the same aircraft to talk to each other.

Unlike the situation just two years before, many airlines were convinced that they could navigate using only Omega navigation equipment. Our years of marketing the Omega System, not just our product, were beginning to pay off. The decisions by Pan Am, KLM, and TWA to replace their LORAN-A equipment with Omega instead of the more expensive Inertial Navigation Systems also helped our cause.

The regulatory agencies require at least two navigation systems on an aircraft flying oceanic routes, which are out of range of ground radars. Today airplane operators preprogram entire route structures into a navigation system's computer memory, but in 1977 avionics technology was such that pilots had to insert the latitude and longitude of the first nine waypoints. To enter a waypoint, a pilot had to make up to 17 keystrokes on the system's control and display unit. For nine waypoints, that was 153 keystrokes. For two systems the pilot had to make 300 entries to program the system for the first section of the flight.

The airlines wanted Omega Systems that could communicate with each other so that a pilot could electronically copy waypoints he had entered into one system. Another reason for that digital link was to prepare for the day when pilots could use Omega and Inertial navigation systems together. Both systems had advantages and a combined Omega–Inertial Navigation System installation had many benefits. That was to come later, and again Canadian Marconi pioneered the concept.

Our engineers were taking care of our first customers, Pan Am and Boeing, helping the Production Department, and designing the new CMA-771 system. They were also solving several problems that were cropping up in our many customer flight tests. Some of us were also working on the airline committee that was writing the new Omega System specification.

All that work helped our current business and future product development, but what we needed was a future. What we needed was a real marketing department. I was still a one-man show for our airline, airframe manufacturer, and military markets, although Peter Gasser somehow found the time, while managing our engineering team, to spend a lot of his time marketing. Carl Hanes, supported by Tom Kelly, was still doing yeoman service developing a market for our CMA-734 system in the business aviation market. However, it was up to us to capitalize on our Pan Am and Boeing contracts. Then was not the time to sit back and congratulate ourselves.

Soon after the Pan Am contract, relief arrived. Henry Schlachta joined our small Omega marketing team. Henry came from a group that was starting to launch a telex exchange system, another product line championed by Don Mactaggart.

One might wonder what the Avionics Division was doing in the telex business. It was another bold venture supported by Kieth Glegg when he was vice president. The company was trying to diversify, and as so often happened, the Avionics Division took the lead.

While developing new avionics equipment, we had learned a lot about microprocessors and digital electronics techniques. Aircraft had many processors, many of them communicating with each other. That was one form of distributed processing. Don championed his idea to use several small computers, instead of one large computer, in a telex exchange. He was convinced that a telex exchange that used distributed processing, with several small computers, would be more economical, reliable, and flexible. In the telex exchange industry, distributed processing, such as Canadian Marconi was proposing, was a new idea, but the idea gained acceptance.

Canadian Marconi parlayed that acceptance into the largest international telex exchange contract that a communications service company had ever awarded. Our contract was with British Telecom. Our competitors had been the telex industry giants. We were newcomers to that industry—it was a remarkable achievement.

John Simons, before he became vice president, had been the group manager for that venture. The telex group was still finalizing the design of the system for British Telecom and manufacturing its parts for that huge contract. At the time it did not need more business, and so it agreed to release Henry Schlachta to a team that desperately needed more help and more business.

Henry is a seasoned marketing professional. Standing about 5 feet 10 inches tall, with a full beard, he could masquerade as a Santa Claus, though a friendly disposition alone does not sell complex products into a complex market. Should he read this book, I hope that he will excuse me if I describe him as a streetwise marketing person. I use that phrase as a compliment.

I believe in marketing people who have a gut-feeling for a situation. Such intuition only comes with talent and experience. Henry has both. Frequently, in a high-technology business environment, the most difficult task for marketing people is selling their ideas in house. Customers are often more receptive than one's own colleagues. I did not always agree with, or understand, all of Henry's intuitions. One such disagreement cost us a potentially lucrative U.S. Army contract in the early 1980s. Henry thought he knew what the winning price would be. He based his opinion more on intuition than solid marketing information. At least, that was my judgment. I disagreed with Henry, but he was right, although it was questionable whether we would have made a profit at the price he was suggesting. Sometimes, however, profits have to wait until a company has established a strong position in a particular market segment. When Henry joined the Omega team in 1977, he quickly became, and remained, a valuable member of the Omega marketing team.

By that time Carl Hanes was marketing full time to the general and business aviation markets. He had officially transferred out of the Avionics Product Support Group where he had been for many years. In addition to general aviation marketing, Carl provided our essential link with the various regulatory agencies around the world. He helped us to understand and to respond to the many regulations with which we and our customers had to conform. He helped us to prepare the large amount of paperwork that we had to submit to those agencies. Also, Carl worked with the agencies when they were writing many of their specifications. That helped us to keep abreast of and to anticipate regulations, which are essential elements of any product campaign, although accountants cannot always record the rewards of such work in a company's income statement.

Carl performed that service until the day of his retirement in 1990, some 13 years later. The cheerful Irishman, Tom Kelly, who remained in product support, continued to help Carl. When Tom reached 65 years of age, we gave him a set of golf clubs, but he used them only on weekends. Tom was not about to become a full-time golfer. He shunned retirement and returned to work the day after he was given the clubs. He was having too much fun to retire, and we were glad to have his experience available to us.

During 1976 and 1977 Carl and Tom signed up 5 Canadian and 19 U.S. general aviation fixed base operator dealers. Most had been active on our behalf, and they had placed orders for several of our CMA-734 systems, the world's smallest and least expensive airborne Omega Navigation System. To produce those sales, our general aviation dealers had done a superb job in supporting our marketing effort. Carl and Tom, of course, had played a significant role in securing that modest start to our general aviation business. They had pursued every one of the some 100 orders that we had received by the end of 1977. Most of the orders were for only one or two systems. Only those with persistence succeed in this market.

Carl had to appease those customers whose equipment deliveries were delayed as we converted CMA-734 systems into CMA-740 systems for Pan Am. Fortunately he succeeded, and we did not lose too many customers due to our decision to give priority to Pan Am.

It was not a decision that we made easily. We resolved that in the future we would do a better job in meeting our commitments. One way we accomplished that resolution was to build more systems than we had orders for.

Companies do not make money with products on the shelf, just as airlines do not make money with aircraft on the ground. It is also unprofitable, to spend money prematurely. The key to success is to understand the market requirements for product delivery lead time, the time between receiving an order and the delivery of the product. An accurate and honest assessment of future sales is another part of the calculation for just-in-time inventories. Only in that way can a company hope to minimize its product inventory.

The markets that we were addressing had a complex set of product delivery lead-time requirements. General aviation companies frequently wanted a product within days or a week of placing an order. The airlines would often take more than a year to decide whose product to buy. Then they usually accepted a six-month delivery schedule. Even that lead time could be a challenge because our product manufacturing cycle averaged nine months. We had a three-month guessing period. However, at that stage in the Omega business the airlines that were buying the system were also rushing to meet near-term deadlines, and so they required a much shorter delivery schedule.

The military services remained the most predictable. In peacetime their procurement activity usually takes place over several years. Once they have made a decision, they give their suppliers at least a year before they have to start delivering products.

We had to continuously reassess the market situation, not only to minimize our inventories, but also to make sure that we had Omega Systems available for short lead-time orders. In our complex market situation, just-in-time inventory management required innovative procedures and a team effort. The marketing, engineering, purchasing, contracts, and production departments worked closely together to speed up or slow down production to meet our market projections. When we released a production order for a new build, we first would order the parts that we needed from outside vendors. Once we received those parts, we were in control of our production. Then we could slow down or speed up production with relative ease. Our ability to deliver systems on short notice and often off the shelf, substantially increased our order bookings.

I do not know how many trips Carl Hanes and Tom Kelly made in 1977 to business aircraft operators and our installation dealers. I do know that they were rarely in Montreal. By the end of 1977, customers had installed our general aviation Omega System into more than 20 different types of fixed-wing airplanes and helicopters. That was an accomplishment because each new aircraft type presented its own unique Omega installation problems. At those times our fixed base operator dealers, which were doing the installations, and our field engineers proved their considerable commitment to our mutual customers.

Our field engineers had to help check out each of those early installations. In time, the fixed base operators became quite expert in installing Omega Systems. Then they did not need our support quite so often. Initially, though, we had to make a considerable investment in the field support of our Omega products. It was a necessary and worthwhile investment. Our support did not end with installation support. Our Training Department had to train all our customers. Even if the order was for one system, there were several pilots that needed to be taught how to operate the system.

We preferred to train our customer's pilots personally rather than rely on impersonal audiovisual training equipment. Sometimes one of our continuously traveling field engineers, Ormee Chamberlain, Bill Lovett, or George Walford, would provide in-the-cockpit training. However, usually the training task fell to Pierre Fournier or his manager, Russell Kelly, both of whom were committed to the task of teaching people. We used every opportunity to maintain a personal contact with our customers. Only then could we develop the personal relationships and market knowledge necessary to remain successful.

Pierre Fournier and I developed a special comradeship over the years. Often our conversations would include some remark about me being a goddamn limey. Then I was obliged to respond with some similarly disparaging remark about Peter's heritage. But our relationship was built on mutual respect.

Over time Pierre held training courses for our customers in more than 50 countries. Rarely did our customers come to us for training, and onsite support rapidly became our hallmark. Many of the larger firms expected their customers to come to them for training and other types of help. They missed an opportunity to understand their customers. Companies can only gain that understanding by sending as many employees as possible to work with customers in their own territories.

In addition to our individual business aircraft operator sales, Carl and Tom secured business with several business aircraft manufacturers. The U.S. subsidiary of France's Falcon Jet Corporation, Gulfstream, Learjet, Beech Aircraft, and De Havilland of Canada were all customers.

Before we had completed our negotiations with Pan Am early in 1977, I made a third visit to my favorite non-English-speaking country—Brazil. Once again, Antonio Miranda and I visited Varig. This time the Varig personnel showed considerable interest in using our CMA-740 system in their B-707s. We appeared close to finally doing business with Varig.

During the visit I persuaded Assis Arantes to visit Pan Am. I thought it would be useful if Varig discussed the Pan Am flight test results directly with Pan Am, and Assis agreed. I was confident that Varig would get a favorable report on our system from Pan Am. I also knew that Varig would hear favorable remarks about the Bendix system because it also had performed well in the Pan Am flight tests. However, I have found that prospective customers respond more favorably if you constructively help them in their evaluation process.

Many manufacturers have lost both credibility and customers by adopting a marketing strategy that focuses on denouncing the competition. We considered such marketing tactics to be those of the desperate. Marketing should emphasize the product's and company's strong points, as well as any unique aspects of the entire commercial proposal. Suppliers should let the customers compare the products based on the information they have been given or product demonstrations. In this case we did not even have any basis for critiquing the Bendix system.

A few weeks after my visit to Brazil, Assis Arantes travelled to New York. There he met with Jacques Raia and Pat Reynolds at the Pan Am base at the John F. Kennedy International Airport. By the time of the visit, Pan Am was willing to admit that its choice was now between the Bendix Corporation and Canadian Marconi. I went to New York to have dinner with Assis Arantes. We tried to use any opportunity to remain close to our customers or the prospective ones. The next day Assis and I traveled by taxi together to Pan Am. I declined an invitation to sit in on at least part of their meeting. It was a time to be discreet because nobody likes an overzealous sales person.

Assis was encouraged by what Jacques and Pat told him during the meeting. During our taxi ride back to downtown New York, Assis asked me if we would support a Varig flight test program. Naturally, I readily agreed, though we had a full slate of evaluations already under way. Assis knew that no Omega manufacturer had a design that could handle all the known Omega signal problems. Also, he knew that as operators flew more Omega Systems, to more places, we would uncover new problems. And so he asked me if we would guarantee the navigation performance of our system.

This was the first time that a prospective customer had asked us to consider a contractually binding performance guarantee. It was a difficult question to answer. I knew we still had much to learn about the behavior of the Omega signals in all parts of the world and at different times. I did not know what problems we would find on the Varig route structure, even though its test of our first Omega System, three years earlier, had shown that navigation to within 4 nautical miles was possible. Assis wanted an answer to his question, and I wanted to give him a figure that I was confident we could achieve. After some discussion about the potential for unknown problems, I answered Assis' question with something like this: "Assis, I am sure our system will navigate to within 4 nautical miles, 95 percent of the time, on Varig's routes. If it doesn't, you have my promise that we will make whatever modifications to the equipment that are necessary to meet 4 nautical miles."

I was serious about that commitment, and I knew I would have the backing of my colleagues on the team. It is necessary for any marketing organization to be and to act as part of a product's overall team. Marketing personnel also should only make promises, verbal or otherwise, that the product team can and will support.

What I did not know when I made that promise was that two years later and after Varig had been using our system for a year, we would face the worst Omega signal problem anywhere. It took us 12 months and a considerable investment to solve the problem—but that is a story for another chapter. Varig rewarded us for our commitment to our promise. Twenty years later it is still a Canadian Marconi customer.

During my third visit to Brazil, early in 1977, Antonio Miranda and I also visited the Brazilian Air Force and Embraer, Brazil's aircraft manufacturer. Embraer was completing the design of a small twin turboprop transport aircraft that it called the Banderante. The Brazilian Air Force had ordered several, and Embraer expected to sell the aircraft to the Chilean Navy and other foreign operators. Eventually, the Banderante became a successful military transport and maritime patrol aircraft, as well as a commercial passenger aircraft for short routes. Antonio had persuaded the Brazilian Air Force to buy our CMA-740 system to test on one of the new Banderante aircraft.

The Brazilian Air Force, unlike most of our other customers, could not accept a system free on-loan. The order, which we received before the close of 1977, was useful for our marketing. We could boast about the Brazilian Air Force being a customer. We did not have to admit that it had bought only one system. Like most sales, regardless of size, it led to the sale of many more systems to both the Brazilian Air Force and Embraer.

Antonio and I were working well together as a team, and we were also friends. Our goal was to capture the entire Brazilian Omega market—we succeeded.

Varig started flight testing the CMA-740 in May 1977, and the trials lasted well into 1978. From the perspective of a search for excellence, I consider the Varig story to be almost as significant as the Pan Am story. Consequently, I refer to our business with Varig in several places in this book. We learned many important lessons from our dealings with Varig. Varig helped to improve our understanding and skills in negotiating and managing international commercial airline contracts. Our association with Varig also fully tested our commitment to product support.

We had our launch customer for the civilian airline market, and next we needed to secure as broad a customer base as possible. With a broad customer base, we would have long-term security. Dynell had a foothold in Europe with its KLM contract. The Bendix Corporation had TWA as its launch customer. The powerful Litton Aero Products Division also had an Omega System in production. We had strong market prospects in South America with Varig and Lan Chile. We were confident that my visits to Japan would eventually lead to a contract with Japan Airlines. We viewed it as our prospective launch customer for the East Asia market. We needed to work farther south in the Pacific, and we had to find a way to penetrate Europe.

To help our European effort we drew up a new agreement with Marconi Avionics in England. This time instead of dividing the world in half—a marketing strategy that had not worked with our first-generation Omega System—we agreed that it would have exclusive rights in England and nonexclusive rights throughout the rest of the world. That may not sound like an improvement on our first agreement, but it worked much better. We agreed to treat each customer on a case-by-case basis.

Because we owned the product, the agreement was a little one-sided. If Marconi Avionics identified a potential customer, it would immediately advise us. We would then decide which company had the best chance of success. If we identified a prospective customer in a region where we thought Marconi Avionics was stronger, we would offer to let it pursue the prospect. Sometimes prospective customers would request a proposal for a complete communication and navigation systems installation. Then the company that had the most expertise, or products, in that category would take the lead. Sometimes we identified a customer in a territory where one company had a better reputation and the right contacts. We had very few disagreements.

We continued to manufacture the product in Canada, though we would have let Marconi Avionics manufacture it for a large enough contract and when it was politically expedient. We transferred the products at prices it could markup to cover its own expenses and profit and sell at the same prices we were quoting. It was a simple but effective agreement. Unlike the first agreement this involvement with Marconi Avionics was voluntary and on a case-by-case basis. That company was more motivated because it was only pursuing those customers that made sense for its situation. With this arrangement, we also had control of our marketing throughout the global market. In the first arrangement, we controlled only half of the worldwide marketing effort, even though it was our product and our future at stake.

The market that we did not have adequately covered was the North American domestic airline market. It was meant to be the easiest market for us to penetrate. It was probably the hardest. In that market, we had the strongest competition, and so most sales were the result of a classic head-to-head competition. We did not have any strategic advantage. In fact, in the domestic U.S. market, we were definitely the underdogs.

The domestic carriers conducted their business operations differently than did international carriers like Pan Am. At first we did not understand the subtle differences, and we paid a penalty with lost sales. Our most important observation was that many of them appeared nervous about dealing with a foreign supplier. In 1977 they could survive, even prosper, with that attitude. Today we all must learn to conduct our business in a global market. Communications, trade agreements, and the shift toward a global economy are just some of the reasons why we must think and act globally if we expect to succeed. Another observation we made was that many of the domestic carriers had developed strong product and company loyalties with their U.S. suppliers.

Soon after we received the Pan Am contract in April 1977, Eastern Airlines announced plans to buy Omega Systems for its Lockheed L-1011 airplanes. The inevitable flight evaluation was soon under way. Litton, Bendix, Tracor, and Canadian Marconi provided equipment for the tests.

Litton carried out the installation survey and chose the all-important location for the Omega System's antenna. When it was our turn to fly, we immediately began to have problems. It did not take long for Robert Baillie, our installation expert, to find out that another system was interfering with ours through the antenna. Eastern would not move the antenna because that was expensive and time-consuming. Moreover, the Litton system had worked. As a result, once again several of our engineers were taken off other work to design a filter that would block the interference. That project came when our priority had to be the preparation of the equipment for Pan Am. We were unable to find a solution in time, and so our system did not perform as well as the Litton system in the Eastern flight tests.

Also, we were battling other more insurmountable market forces. Many airlines still considered Canadian Marconi as a newcomer to their market. Litton and Bendix had sold, and continued to sell, avionics to many of the U.S. domestic air carriers and several foreign airlines. They were known suppliers with an in-country product support infrastructure. We were on the other side of that friendly but nonetheless national boundary.

Late in 1977 Eastern issued its Request for a Proposal, which included a requirement for a very short delivery schedule. Peter Gasser and I agonized for days about how, and even whether, we should respond. We knew that we could solve the interference problem on the Eastern L-1011s. What worried Peter most was that we did not have the resources to support both the Pan Am and Eastern delivery schedules. To meet the Eastern schedules we would have had to delay some shipments to Pan Am, but we were not prepared to renege on our commitment. Pan Am deserved better treatment than that, having given us our big chance.

To counter Litton's advantage as a U.S.-based company, we felt that we would have to invest in a Miami-based repair center. We knew intuitively that we would have to make that kind of investment to secure a major U.S. domestic carrier as a customer. We did not know what Litton would bid. To have any chance of success, our price would have to be lower than Litton's. It was the wrong time for us to make that kind of investment. We also had our commitment to Pan Am to consider. Of course, we could have lied to Eastern about our ability to deliver to its schedule, but that was not our style.

Peter and I weighed all the factors realistically and honestly. We came to the conclusion that Eastern's contract was one that we could not win. Reluctantly, we advised Eastern that we were not going to bid. It was the first time that I had participated in a comprehensive analysis to determine whether or not we should bid on a particular job.

The bid or no-bid decision is one of the most difficult decisions any supplier has to make. Usually, those decisions need to be made only for multimillion dollar procurement activities, those where the preparation of the proposal itself may cost tens of thousands of dollars and where there is an investment risk. Occasionally, a company must carry out a bid or no-bid analysis even for small orders. Such was the case we faced with the Eastern Airlines Request for a Proposal.

Regardless of the size of the potential order, the bid or no-bid analysis is the same. Companies must ask themselves and honestly answer many questions. Do we have the experience and technical capability for this job? Are our best engineers, facilities, and subcontractors available? Do we understand the requirements? Have we worked with the customer? What are our strengths and weaknesses compared with those of our competitors? Can we meet the specification? Do we have credible cost information and a pricing strategy that will not bankrupt the company but still win the job? A company making a bid or no-bid decision must answer other questions, but those are the main ones.

Our decision not to bid surprised Eastern Airlines. Had we misjudged the situation? Perhaps we had, but based on the information that we had available, I do not think so. Early in 1978 Eastern awarded Litton the contract as we had predicted. Litton had its launch customer. It was that fact, more than not winning the contract, that concerned us the most. However, we retained our integrity with Pan Am, and through it our integrity with many other airlines. It would have had aircraft idle on the ground if we had not fulfilled our delivery commitments.

Shortly after we withdrew from the Eastern competition, its schedule slipped so much that we could have satisfied it without jeopardizing our deliveries to Pan Am. However, we did not know that would happen at the time of our withdrawal. Later in 1978 Litton once again beat us by capitalizing on its existing customer base and securing a contract with American Airlines. Litton was starting to gain a considerable foothold in the U.S. market.

By the end of 1978 Litton had emerged as our strongest competitor. It maintained that position for the next 15 years, as we fought for and ultimately shared most of the worldwide airline Omega Navigation System market.

When Henry Schlachta joined the Omega marketing team, he began to share the airline marketing with me and the general aviation marketing with Carl Hanes. We developed a strategic marketing plan that had some basis for success. We had the beginnings of a customer base, and we were starting to understand the nuances of the worldwide market for our product. Our previous strategic plans were actually tactical plans. They included market data and near-term action plans. What our plans lacked was a long-term strategy backed by accurate and specific market knowledge. A strategic plan for a particular product line can only be formulated once you have experience in that product's market. Anything else is just guesswork. Strategically, we knew now what we had to do and what was within our capabilities.

We had to build a general aviation customer base with an emphasis in the United States. We had to capitalize on our Pan Am contract, particularly in South America and the Pacific. We needed a launch customer for Europe and from it build a European market. We had to find a way to overcome Litton's successes in the United States, because we needed a U.S. civilian market. Rather than put all our efforts into the major domestic carriers, where Litton was strongest, we concluded that we should place more emphasis on the smaller carriers. We knew that the FAA would soon publish guidelines for certifying Omega Systems for use within the domestic airspace. We no longer would have to consider the oceanic operators as our only customers. The FAA issued those guidelines in an Advisory Circular at the end of 1977.

We decided that the African and Middle Eastern markets would have to wait, except for those customers that we would secure through Boeing. Also, we realized that we did not have the resources to launch a successful marketing campaign in Eastern Europe, India, or China. We would react to market opportunities in those regions, but we would not take the initiative.

Our strategic plan included building on our alliances with Marconi Avionics in England and Marconi Italiana, as well as the major airframe manufacturers.

We included the military markets in our strategic plan for the Omega product line. However, they were no longer the dominant market. We wanted to achieve a balanced global military and civilian market mix to have a sustainable long-term business.

We could reasonably support a long-term strategy. Our objective was to maintain a 10 percent or better sales growth each year for 10 years to reach $10 million in annual sales. We reached our goal in eight years. Our principal strategic goal was to build a broad worldwide market. Only then could we expect to build a business with new commercial avionics.

A $10 million annual sales goal was a reasonable expectation. We recognized that the market for navigation systems was limited. We did not expect to maintain a 10 percent annual growth rate after we had reached our $10 million target. What we did expect was to leverage our Omega success into civilian market sales of new avionics while maintaining a modest Omega business.

Our plans were bold considering that we were a marketing department of only three people. However, by developing a team approach to the business, we could augment our front line marketing with personnel from the engineering and product support parts of our team. We could also rely on our contracts administration staff to help in the field or with follow-on proposal work.

The ability to call on that type of help in marketing is one of the many advantages of having a cohesive team for a business venture. Though we centered our team around the product manager, there are other ways to achieve a cohesive team and entrepreneurial spirit, even in a large firm. The key is to create companies within companies wherever possible without duplicating common functions, such as production. Despite the support that we had, it was a daunting prospect for three people to be thinking about tackling the three major sectors of the aeronautical market on a global scale.

Henry focused mostly on the U.S. and European civilian markets., leaving me free to pursue opportunities in South America and Asia. I also covered the large airframe manufacturers and military agencies around the world.

Unfairly, we listened to Henry with some skepticism when he returned from a trip to Denver. He told us that he was close to a deal with a travel club. Most of us had not heard of travel clubs and certainly not one called Ports of Call Travel Club. The first question was from our comptroller was, "What is their credit rating?"

That organization owned two old Boeing B-720 airplanes. It had dues-paying members from throughout the United States and offered them inexpensive vacations in many countries. Despite our initial skepticism, Ports of Call did buy our Omega Systems, and the club became an important strategic customer. Other travel clubs followed its lead.

More important, in October 1978, Ports of Call became the first commercial operator the FAA approved to use Omega equipment as the primary means of long-range navigation. The approval included all of the club's worldwide routes. For its certification request the club had collected, for many months, navigation data from the two Canadian Marconi Omega Systems on its aircraft. Ours was the first Omega Navigation System approved for use as the sole means of navigation. It was a strategic coup for Canadian Marconi.

The day before Christmas 1977 Henry received a call from another airline that he had been pursuing, British West Indian Airways in Trinidad. It asked him if he would travel to Trinidad the day after Christmas because it wanted to finalize a contract. As was typical of our team, Henry did not try to delay the meeting but left his family's post-Christmas reunion and flew to Trinidad. He finalized the contract during the visit and came back to work, at the start of the New Year, justifiably proud of his efforts.

For many years the Lockheed C-130 military transport aircraft had used the Canadian Marconi Doppler Navigation System. Eventually the time came to upgrade the navigation system on the aircraft. The U.S. Air Force had selected the Dynell system for its C-130s. However, Lockheed decided that for its foreign military customers it should offer an Omega and an Inertial Navigation System combination. It selected Litton as the supplier for the Inertial Navigation System. Then, in 1977, it held a competition to select an Omega System supplier. Because it was providing the inertial system, Litton was in a strong position for the Omega competition.

Once again we ignored our internal job descriptions and used the best people we had to help us compete for that important contract. Lionel Leveille, the manager of our Avionics Product Support Group, who later held the top job in our division, had dealt with Lockheed-Marietta for many years during our role as the Doppler system supplier. But those days were coming to an end. I asked for Lionel's help, which he gave readily. We made many trips to Lockheed because it was essential that we understood its special requirements and needs; we listened. Lockheed had some concerns about using the Omega System. To some people, the system was still unproven. We did not have years of operational experience with the system, and that prompted questions about its reliability and operating costs.

We won the contract, but only after Lionel had prepared several cost of ownership and support analyses for Lockheed and after we had put together a proposal that addressed all its requirements. Winning that contract was a major coup for us. We had secured the type of contract that everyone in the aeronautical business strives to obtain—a contract that produces steady long-term business. We were secure in the knowledge that every Lockheed C-130 sale to a non-U.S. customer would include our Omega System. Because the C-130 was still in demand throughout the world, we were assured of at least a modest business base. We needed such a base to continue in the very competitive and, for us, risky civilian business. We were not expecting the Lockheed contract or our other military contracts to cover civilian contract losses because all our bids were profitable. What we needed was the assurance of continuous production because, if the cycle is broken, costs escalate, forcing a company into an uncompetitive situation.

A secondary, but important, aspect of the Lockheed contract was that we had beaten Litton. We had beaten it with a major customer in its own country, when it was the supplier of the other navigation system on the C-130. However, once Lockheed had made its decision, we put competitive considerations aside and worked to ensure that our two products would perform together as Lockheed, our mutual customer, expected.

That cooperative effort was not the first time nor was it the last time that we worked with a competitor to support a mutual customer. Only the foolish and inexperienced do not know when to compete and when to work together.

Marconi Avionics had helped us to secure two customers in England. Like most beginnings, it was modest, but sales gave us encouragement because they were one more step toward broadening our market base.

The declaration of the international 200-mile coastal zone agreement created a new opportunity for aircraft manufacturers. We had identified that opportunity and had pursued it in several countries. All countries with coastlines were now candidates for patrol aircraft. Mostly, those countries wanted to prevent foreign vessels from fishing in their waters, a practice that by that time was illegal by international agreement. To prosecute, countries had to prove the offense with photographs and irrefutable evidence that the ship was within their 200-mile zone. Producing that evidence required a navigation system that remained accurate after many hours of patrol. The Omega System is uniquely suited to that role.

England's Hawker Siddeley Aircraft decided to produce a version of its HS-748 turboprop aircraft for coastal patrol. De Havilland of Canada offered a version of its Twin Otter aircraft. Brazil's Embraer offered its Banderante, and Aeritalia in Italy offered its G-222. Those aircraft cost much less than the standard military maritime patrol aircraft. Boeing offered a patrol airplane based on its B-707. However, it withdrew from this market soon after the Canadian Armed Forces selected the world's maritime patrol workhorse, the Lockheed P-3.

Marconi Avionics helped us to secure a contract for Hawker Siddeley's HS-748 Coastguarder. All we needed was for the company to sell some airplanes. Despite much initial optimism, it did not sell any Coastguarders. In fact, the maritime patrol market did not produce a lot of new aircraft orders, as many of us had hoped. Nevertheless, the contract with Hawker Siddeley and our loan of a flight demonstration system provided us with a useful marketing story. Also, Hawker Siddeley helped us to better understand the unique navigation requirements of that market. We designed special features, such as an automatic search pattern mode, for our system. The credibility that we gained through that association and the special features that we designed helped us to secure other contracts for similar applications.

Marconi Avionics also helped us to secure a contract with Monarch Airlines. It was our first European customer, and the contract gave us the incentive to set up Marconi Avionics as our first European repair center.

During 1977 we had to look for and seize every opportunity. It was a time to be bold. We had a start in the market. Success only comes to those who want to succeed and who earn their success with hard work.

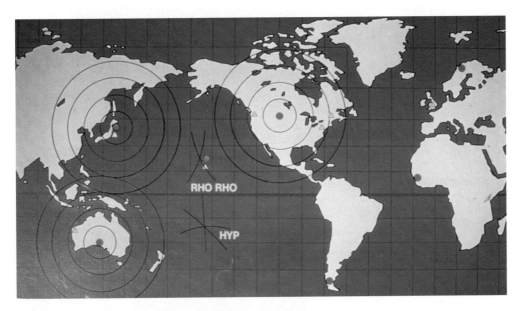

Positioning fixing with the Omega network

An early control display unit of a Canadian Marconi Company Omega System.

Alphanumeric control display unit.

Canadian Marconi Company Omega CMA-719 receiver processor unit.

CMA-740 receiver processor unit.

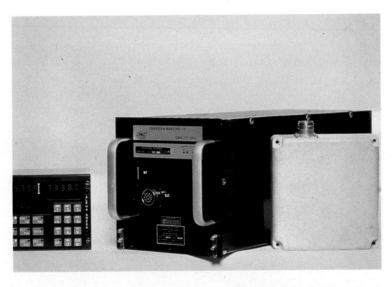

Canadian Marconi Company's CMA-771 Omega Navigation System.

CMA-900.

CMA-900 FMU with antenna and GPS board.

12

Building Our Market

*The heights by great men
reached and kept
Were not attained
by sudden flight,
But they, while their
companions slept,
Were toiling upward
in the night.*
—Henry Wadsworth Longfellow

By the end of 1977 we had completed the delivery of 105 Omega Systems to Pan Am. We had also started deliveries to Boeing for several of their B-737 and B-727 customers. We had secured a contract with Lockheed Georgia as the exclusive Omega supplier for C-130 transport airplanes it sold to non-U.S. operators. Also, we could boast Hawker Siddeley Aircraft, in England, as a customer, although orders would only come if it succeeded in selling its Coastguarder aircraft. We also could claim the Brazilian Air Force as a customer.

Henry Schlachta had convinced De Havilland Aircraft of Canada to offer our Omega equipment as an option on its Twin Otter aircraft. One of our first customers through De Havilland was Greenlandair, and so we could add one more customer to our airline customer resume.

Through Boeing, we had Liberian Airways and the Niger National Air Squadron as customers. Because Boeing was offering our system only as an option, I had to go to Boeing frequently to help sell our system to a visiting customer.

When representatives of the Liberian government were visiting Boeing to negotiate the purchase of a B-737, they asked to meet with the supplier of the Omega equipment. With only a few hours notice, I left for the Boeing plant in Seattle. We did not ignore any opportunity to meet with a potential customer, no matter how inconvenient.

Immediately upon my arrival at Boeing, I was escorted to the conference room to meet the Liberian negotiating team. Their leader, a tall large imposing man with a commanding voice, immediately wanted to know why they should pay extra for the Omega equipment. I looked to my Boeing colleagues, whose expressions suggested that I should be the one to answer the question. Slowly I explained that Boeing equipped the basic B-737 for domestic routes where ground-based navigation aids were usually available. Now my Boeing colleagues helped me. As delicately as we could, we pointed out that because the president of Liberia, as well as their airline, would be using the airplane, its destinations could be anywhere in Africa or Europe.

When we produced route charts for Africa, they could see that they would not be in range of ground-based navigation stations all the time. Then they would have to rely on navigation equipment that did not need local ground stations. The Omega System provided such a navigation service; the alternative was an Inertial Navigation System at nearly four times the cost of the Omega equipment.

That short anecdote helps to show that to be successful a company must understand the environment in which its customers operate and how they will use the equipment. The anecdote also shows that companies cannot take any sale for granted. A company must market every step of the way to the contract, whether it is through a third party or directly with an end user. Successful companies never assume that the next link in the marketing chain is the only place where they have to sell. Dealers, and other part-time representatives, in particular, need lots of support.

Marketing continues even after a contract is signed. Some people will go so far as to say that at that point marketing really begins: it is called product support, which does not mean only repairing the equipment when it fails. Knowledgeable companies make sure that their customers are satisfied with their product and their service because satisfied customers come back for more.

As usual, we had several prospective customer evaluations in progress. During 1977 I made three visits to AeroMexico in Mexico City to discuss installing Omega Systems in its DC-8 aircraft. The airline was now flight testing our CMA-740 and the competing products from Litton, Tracor, Bendix, and Dynell.

Varig had been flying our equipment since May on routes that covered South America, the Pacific, and the South and North Atlantic from Rio de Janeiro to New York. It was the first time that any operator had flown Omega equipment on some of those routes. Once again, we found ourselves involved in pioneering work.

Occasionally, on some of those routes, the system would have large navigation errors. Then one of our field engineers had to fly with the system to find out what was going wrong. It was becoming clear to us that the Omega signals did not always behave as the theory predicted they should. We had no choice but to assign two engineers full time to the task of analyzing the flight data. One of those engineers was Robert Baillie, who had dealt with similar unusual problems in the past.

We discussed our findings with Peter Morris, a scientist at the U.S. Coast Guard Omega Navigation System Operations Detail. Peter, one of the world's foremost authorities on Omega signal propagation, helped us to understand what was happening. Once again our engineers changed the software.

We knew that Varig, an important customer in its own right, also would be a strategically important customer for the South American market. We needed to win the Varig competition, and so we gave it priority over most of the other opportunities we were pursuing.

Somehow we found the resources, time, and energy to support flight tests with six other operators. One was an evaluation by Britain's Royal Air Force. It had tested our first system on two occasions during the last five years, and finally it was interested in buying Omega Systems, not just testing them. We hoped that our earlier association with the Royal Air Force would give us an advantage over our strongest rival, Litton—again.

During the autumn of 1977 I made a trip through Asia and the South Pacific. That resulted in requests by Japan Airlines, Cathay Pacific, and Air Pacific for more evaluations in 1978. The trip also gave me an opportunity to visit my parents, who had settled in New Zealand. Air New Zealand and the country's air force were on my list of prospective customers. I confess to a personal motive for wanting to succeed in the New Zealand market. Within 12 months we sold products to both organizations. My parents were happy that it took more than one visit for me to secure those contracts and to ensure that our customers remained satisfied.

Only a few readers will have heard of Air Pacific. It is a small airline, which at the time of our first association flew from its base of operations on the island of Fiji to New Zealand and Australia. It expected to buy only three or four systems; nevertheless, like the large carriers, it wanted to try one first. We agreed to the evaluation because we knew that this was the only way that airline would become a customer.

Under normal circumstances a potential contract for less than $100,000 did not justify the cost of an evaluation. Regardless, we felt that a contract with Air Pacific would influence operators on other Pacific islands. Also, we were still trying to build a customer base and with it our credibility as a bona fide commercial avionics supplier. The only way to achieve that goal was to invest in marketing activities such as customer evaluations.

The larger corporations like Bendix and Litton already had a customer base through their other avionics products, whereas we were starting from zero. One of our strengths was that we understood that we had to be more accommodating than our competitors. For us, building a customer base was as important as revenue. Our flexibility showed in many ways, whether it was supporting product tests by a small airline halfway around the world or designing special product features for a particular customer. Our willingness to be flexible served us well over the years.

We noticed that many of the larger corporations separated the different disciplines required to bring a product to market. As the firm grew, so did the bureaucracies governing the various operating functions. Perhaps the individuals in those organizations were not willing to exert the extra effort required to champion a special activity in that type of organizational structure. The result was a more rigid approach to customers. Some companies even gave the impression that they only wanted to sell what they had designed. Few companies can adopt and succeed with a take-it-or-leave-it attitude. Our sales proved that our product-manager-runs-the-entire-show organization was effective. That organization allowed us to use our initiative and to respond to customers.

As a new entrant into the civilian market, we were encouraged that we were holding our own against powerful competition. When measured by the number of individual customers, our success rate was nearly 50 percent. We were successful partly because we had started early. Also, we had made the long-term investment necessary for success in a new market, and our payoff was a growing worldwide customer base. Ten years later, it was easier for us to market other new commercial avionics products because we had already established ourselves. By then we could concentrate on selling our products because we no longer had to sell the company.

Two military contracts that Litton, not Canadian Marconi, secured were particularly frustrating. We made some serious errors, which we tried not to repeat in the future.

During the Farnborough Air Show, we got our first sign that the Royal Air Force was beginning to favor the Litton Omega System. How could this be? we asked ourselves. We had worked so long with the Royal Air Force that we assumed it would not question our ability—our first mistake. Both the Canadian Marconi and Litton systems had performed well during the flight test program earlier. Both companies had submitted bids, and we were sure they were comparable. Canadian Marconi and Litton had teamed with companies in England, which would provide in-country support to the Royal Air Force. We had teamed with Marconi Avionics. Neither bid contained an offer to manufacture the products in the United Kingdom. License manufacturing had not been a requirement of the procurement specification upon which we had based our proposals.

I now think that not offering to have some parts of our equipment assembled and tested in England, even as a separately priced option, was a strategic mistake. Setting up a second party for license manufacturing is an expensive proposition, and the Royal Air Force procurement quantity did not justify the investment. However, in retrospect, we should have carried out a more comprehensive assessment of the longer-term sales prospects. Then we might have concluded that the investment was worthwhile.

Our problem was that we did not want to win hard enough; perhaps we were even a little complacent. To win, in any competition, you must want to win, but not at any price, and cheating is not an option. In between, though, lie many options available to those with a desire to win, an understanding of the market, and initiative.

The options available depend upon the specific business circumstances. Lowering profit margins, against long-term business opportunities, is one strategy. That was not an option we could entertain on that occasion. We had already reduced our price to a point at which we were running the risk of an unprofitable contract.

A lost leader, more commonly called a buy-in, in a high-technology specialized business is rarely worthwhile. Sometimes there is a temptation to bid that way when the contract has significant long-term strategic importance or when there will be large follow-on contracts. Even then, in the fixed-price contract environment in which we were operating, a buy-in represented a considerable risk. Most of the time a customer will not pay much more for additional equipment, government-calculated inflation increases excepted. Then a company is stuck with a long-term unprofitable contract.

Sometimes it is useful for a company to accept lower profit margins, such as when a company is trying to enter a new market with the view to dominating it. The Japanese have proved the viability of that approach in many markets, which they now dominate because their competitors were not willing to accept lower profits or return-on-investment ratios to stay in the market.

In a globally contested competition, you have to find a way to make your bid more attractive than that of the competition. A sprinter might find a weight-lifting program that gives him the vital edge to be first off the blocks. It does not matter if he wins by only 1000th of a second, he has won.

We did not try hard enough to make our bid to the Royal Air Force sufficiently more attractive or different than our competitors' bids. Offering license manufacturing in the United Kingdom may not have solved our problem, but now we will never know. Without a distinguishing characteristic, our bid was at the mercy of the usual bid evaluation factors.

One of our problems with the Royal Air Force was its perception of the technology we were using. When we first designed our second-generation Omega equipment, single-chip processors suitable for airborne use were not readily available. Therefore, we had designed the equivalent of a powerful single-chip processor by using individual components. When Litton designed its product, some 12 months later, single-chip microprocessors were available. The scientists at the U.K. Ministry of Defence concluded that the Litton system was more powerful than ours because of its microprocessor. They wanted to be sure the equipment they selected had surplus processing power because the Royal Air Force would want new features during the life of the product.

Most customers of an electronics product want one that will easily accommodate their growing needs. The personal computer is a typical example.

The expansion capability of our equipment was considerable. That had been one of the design goals we had imposed on our engineers. We could not afford to design a new product every time a customer wanted a new feature. Our design had to be flexible for economic as well as market reasons. Over the years, we introduced more new features for individual customers than did any of our competitors. In the meantime, though, we had to convince the scientists who were advising the Royal Air Force. We provided technical data to support our claims, and we attended several meetings to try to respond to their doubts.

When customers have developed a perception about a product or a company, it is difficult to persuade them that their assessment is wrong. It is better not to get into that situation, which means a careful and honest analysis to find any weaknesses, real or imaginary, that the prospective customer might identify. A technical proposal has to be accurate, and so it is not always possible to hide a potential weakness. However, there are many ways to describe a product. The technical description should provide sound technical arguments to avoid even the perception of a weakness.

During one of our final meetings with the Ministry of Defence procurement team, the scientists asked us to quantify the excess computing capacity of our system. The right answer was 30 percent, or some similar figure. Instead of uttering two words, our answer was complex. It was technically accurate but left the wrong impression.

Omega Systems carry out many complex calculations. The most complex are those required to correct the received signals for variations in the height of the ionosphere, the Earth's magnetic field, and other variables. We had designed our system to perform those signal correction calculations whenever the system was not performing another navigation calculation. The computer was doing something all the time, even though that was not strictly necessary. That was the long-winded answer to the simple question; "How much spare computing capacity does your system have?"

In my opinion, we lost the competition at that point. We did not really answer the question, and, worse, we led our questioners to believe that our computer did not have any spare capacity.

Answers should be factual, succinct, and pertinent to the question. Elaboration can come in response to follow-on questions. In this case, we should have explained that the system required 70 percent of the available computing capacity to meet its navigation performance specification. We could have stated that for the time being, rather than waste the excess capacity, we were letting the system run additional signal correction calculations.

Marketing people by their nature like to talk. It has taken me 30 years in business to learn to be brief. I am still not succinct enough for some people.

One month after our ill-fated meeting in London, the Ministry of Defence told us that it had selected the Litton Omega equipment. The initial contract was for about 50 systems. Shortly after the Falkland Islands conflict started, I was talking to a colleague who worked for Litton. He told me that it was having to redirect its entire Omega production effort as the Royal Air Force and Royal Navy equipped many more aircraft with Omega Navigation Systems.

We could never have foreseen the Falkland Islands conflict. However, we should have realized that the first contract with such a customer would probably not be the last, despite the initial forecasts. That happened to us with Pan Am, Varig, the Chilean Air Force, and many others. I cannot emphasize enough the importance of first securing a broad customer base. It does not matter if the first individual customer orders are small.

The second military order that Litton won was a personal as well as a business defeat. It was not a large contract, but we did not treat any defeat lightly. The team's strong desire to win drove us, not the fear of reprisal from our corporate management, which continued to support all our endeavors. We were not preoccupied trying to protect our jobs, and so we could focus on our responsibilities.

Soon after our last meeting with the Royal Air Force in England, I made one of my periodic trips through Asia and the South Pacific. We were making good progress with Japan Airlines, and so Japan was my first stop. Then I went to Taiwan to visit China Airlines, which had recently become a customer. It was one of the first airlines to be directly influenced by Pan Am's selection of the Canadian Marconi Omega System.

I was pleased to be returning to the island of Taiwan, although, once again, I would only have time to go to the capital, Taipei, at the northern end of the island. I always found the Taiwanese to be friendly and industrious. Taipei, as I remember it, is a busy place, but the streets are narrow enough to give it a distinctive character. Shops, restaurants, and the smells of spices abound.

After Taiwan, I went to Hong Kong and Singapore. Hong Kong is more crowded than even Tokyo with barely a patch of grass anywhere. Despite the crowds and apparent frenetic pace, Hong Kong is my favorite city in the Orient. It is a shopping and eating paradise, or you can just watch the junks, ferries, and cargo ships plying the water between Victoria and Kowloon. Singapore is a contrast to Hong Kong. For me its wide clean streets and grand buildings do not have the character that makes Hong Kong so special.

The principal reason for my stops in Hong Kong and Singapore was not to sightsee but to interview potential dealers and repair centers. We were anxious to set up a company in the Pacific to hold spare equipment and to support our growing list of customers in that region. I also visited Cathay Pacific while in Singapore, however they had no near-term requirements for Omega equipment.

We did not set up an independent repair center because we developed a better approach by listening to our customers. We offered first Air New Zealand and later Japan Air Lines the opportunity to use their repair shops to repair Marconi Omega equipment being used by other airlines. We also agreed to pay those airlines for work they carried out on equipment that was still under warranty. It was a novel arrangement and one that served us well in several countries.

My next stop was Australia, where my lost baggage from the start of the trip finally caught up with me. There I held meetings with representatives of Qantas and the Royal Australian Air Force.

By now my scuba diving days were almost over, but John Rogers, who had helped me to become a diving instructor, had influenced me once again. He had introduced me to long distance running. After several months of not running more than a block or two, I was now covering 10 miles comfortably. In the upcoming year, 1979, I expected to run my first marathon. I was an enthusiastic runner, and I combined running with sightseeing during business trips. That was how I saw Canberra. The capital of Australia, as I saw it through sweat and with labored breath, is a magnificent planned city with many green spaces. It resembles Washington, DC, but is even more spread out and has newer buildings.

From Canberra I traveled on to New Zealand where I hoped to complete a sale of our Doppler and Omega Navigation Systems with the Royal New Zealand Air Force.

I had worked to develop a viable market in New Zealand, and Air New Zealand was now a customer. I had tried to cover all eventualities. Since my last visit, our New Zealand agent, Eric Pernase, had maintained close contact with our next prospective customer. We visited the Air Force, the Civil Aviation Authority, and other government departments that might be involved in the procurement.

At the end of the visit, I squeezed in a few days with my parents, my sister, and her family. I went sightseeing around Auckland wearing a pair of running shoes. My sister's husband is a world-class sailor, and so during a long weekend we sailed to a group of islands north of Auckland. The uncrowded tranquil coves and grassy islands were a sharp contrast to Asia. I was exhausted; it had been a long trip.

When I left New Zealand, I thought that the Doppler contract was certain and the Omega contract a strong possibility. I stopped off in Fiji for one day to renew contact with Air Pacific; we never let an opportunity pass.

A few weeks later, at an airlines meeting in Seattle, I heard from a Litton colleague that the Royal New Zealand Air Force had, that day, awarded Litton the Omega contract. I was stunned. I immediately telephoned our agent in New Zealand. Yes, the information was correct. The Air Force had told him of its decision just moments before. It had, however, selected the Canadian Marconi Doppler system.

The reasons were twofold. As usual, there was not much difference between ours and Litton's price. The Air Force had decided that it should not have both of the navigation systems in the aircraft manufactured by the same company. It seemed militarily prudent to have redundancy in manufacturers, not just navigation systems. The second reason was the close ties between Britain's and New Zealand's air forces. The former had selected the Litton Omega System, and that choice had swayed the Royal New Zealand Air Force in the same direction. It was a disappointing outcome.

One of the few disadvantages of our product management matrix organization was that our marketing staff was responsible for one product line. We had not yet put into place a multiproduct approach to marketing. I championed that idea and later succeeded in having it adopted without sacrificing the benefits of our product team organization.

The Doppler team was pleased with my efforts in New Zealand, but the Omega team was not. The situation once again demonstrated the importance of securing strategic customers. We had developed a set of strategically important customers—Boeing, Pan Am, and Lockheed—but more would have been better.

The big news for 1978 came during the summer. After four years of continuous marketing to Varig and extensive flight tests, it rewarded us for our efforts. The purchase of our system for its B-707 fleet was the beginning of a long and successful business relationship between Canadian Marconi and Varig. And, we had another strategically important customer.

The story of the final negotiations with Varig is typical for any understaffed organization that is determined to succeed. I tell it not to suggest that I alone secured the Varig contract but to show that a lack of resources is no excuse for failure. Many people on our team were involved at every stage. Our field engineers, Ormee Chamberlain, Don Mackenzie, and Bill Lovett, spent many hours on Varig aircraft. Pierre Fournier, our instructor, almost single-handedly trained all of the Varig flight crews. Gilles Bercier, our senior repair and overhaul technician, helped the Varig maintenance staff set up their Omega repair center and helped them with several system upgrades. Our engineers toiled many hours to make the equipment work on the Varig routes. Our contracts, pricing, and legal staff helped in many ways. Our publications department produced special manuals for Varig, and our production department produced a quality product, all under the watchful eye of our quality assurance department.

However, except for one trip, when our company's legal secretary, Claude Filiatrault, and Barry Howarth, our group manager, accompanied me, I conducted all the technical and commercial negotiations with Varig on my own. Companies who encourage employees to possess several skills will succeed better than those that promote specialization. The expectation of a more volatile global market, the need to retrain people quickly, and the need to improve efficiencies will demand a flexible workforce. It is inefficient to have to send

10 people halfway around the world to negotiate a $1 million contract. For contract negotiations of that size, two people with the necessary skills and knowledge should suffice. Two people at the negotiating table also have the advantage that one can think while the other is talking. There is less chance of making a tactical error if two people work together.

Frequently, at Canadian Marconi, we could not afford to send even two people to negotiate a contract. Most of us on the Omega team who negotiated contracts did so on our own. It is true that we usually spent our evenings on the telephone seeking advice or obtaining more information from the experts. It made for long, tiring days but exhilarating negotiations.

Once Varig had selected the Canadian Marconi Omega System, based on its technical merit, we could focus on the commercial aspects of a contract. In Varig's case, those aspects turned out to be a complex set of documents.

I had two, week-long, face-to-face negotiation sessions with Varig's director of procurement, Arthur Mueller. Like many in southern Brazil, Arthur was of German descent and, like all Varig employees, a professional from whom I learned a lot.

First, we went through the general terms and conditions. Those clauses covered the method of payment, liability, performance default, delivery schedule, taxes, duties, patent indemnity and other terms, all quite straightforward, general, and easy to negotiate. Later, Brazil's monetary regulations complicated our negotiations.

As we negotiated the general terms and conditions, Arthur kept going back to our pricing schedule. However, I had learned long ago to negotiate prices only after both parties had agreed to all the other conditions.

The most difficult part of the negotiations was the product support agreement. In addition to the usual warranties and reliability guarantees, Varig wanted a performance guarantee and other more demanding agreements. A performance guarantee is a reasonable request. A guarantee, though, is only useful to the buyer if the seller is contractually bound to honor the guarantee or face a penalty. When a product relies on outside factors, like the transmitted Omega signals, it is difficult to negotiate a meaningful, but reasonable, guarantee with a penalty. Varig also wanted the performance guarantee to last for as long as it owned the equipment, regardless of where it flew and in which aircraft it installed the equipment. Eventually, we agreed on a set of conditions that satisfied Varig. We took a risk, but as Arthur Mueller repeatedly pointed out, Varig would only buy from a company that would support its products with a binding contractual guarantee.

Varig wanted to repair the equipment even before the warranty expired. Because we had offered a three-year warranty, Arthur and I had to negotiate a special agreement. After much debate, I agreed that we would pay Varig for repairs it carried out on equipment that was covered by the warranty. We had to negotiate a limit of liability. That meant describing the type of repairs that

Teaming a Product and a Global Market

Varig would carry out and agreeing on a set of maximum billable hours for each type of repair. Once again, we were pioneering a new concept.

We were also one of the first companies to enter into third party repair center agreements with selected airline customers. We had already agreed to that arrangement with Japan Airlines and Air New Zealand; now Arthur Mueller and I worked up such an arrangement for Varig.

Because Varig would be performing its own repairs, it needed a stock of spare parts. At that point, we got trapped in the Brazilian export and import regulations. When Varig imported equipment or parts, it took weeks to clear them through customs, and the airline had to pay large import duties. Finally, after consulting with Brazilian government officials, we developed an agreement under which we would consign parts to Varig that it would hold in a bonded warehouse, until needed. When a part was taken from the warehouse Varig paid the Brazilian import duty and paid Canadian Marconi the agreed purchase price of the part. It was a complicated agreement to negotiate because we had to agree on the consignment stock, restocking procedures, and prices. Privately, I had to calculate and consider our costs for that consignment of parts.

At night, in my hotel room, instead of enjoying a Brazilian barbecue at the Rancho Alegre, I was doing overtime with my calculator. I had to estimate the cost of all of the special agreements. Only then could I negotiate the Omega equipment price that Varig would pay. Also I spent many hours on the telephone conferring with my colleagues in Montreal.

It was a multifaceted complex contract, but in the end it satisfied both parties. Assis Arantes, Arthur Mueller, Antonio Miranda, our Brazilian representative, and I celebrated the culmination of our sometimes intense negotiations with a barbecue and several caipirinhas, a potent Brazilian concoction, at Arthur's home.

I judge a successful business relationship if several years after signing the legal documentation I can look back and say it never had to be taken out of the filing cabinet to resolve a conflict. That is just another piece of wisdom that Arthur Mueller of Varig shared with me. Most of our contracts fell into that category.

13

An Organization to Fit the Market

Drive thy business, or it will drive thee.
—Benjamin Franklin

While we pursued airlines, we made progress in the business aviation market. Although Litton and Tracor also competed for that market, Global Navigation Systems Incorporated and the Communications Components Corporation dominated the long-range navigation general aviation market.

Thus we faced a situation where we had a different set of primary competitors in the airline and general aviation markets, which just added to the complexity of our marketing efforts. Nevertheless, Carl Hanes, now helped by Henry Schlachta, began to secure orders for our CMA-734, still the world's smallest Omega Navigation System. By the end of 1977, after only two years of marketing the product, we had sold more than 50 systems. Considering the general aviation market environment and our situation, we were delighted with those sales. For 6 of those 24 months, we had diverted our production to the Pan Am contract. Also, the more than 50 systems represented 50 individual customers. Identifying and selling to those customers had been the work of one person. Also, our small team had handled a multitude of installation design problems.

Our success attracted the attention of a large U.S. avionics distributor, which, late in 1977, suggested that it be our exclusive outlet for general aviation. The distributor had a large sales and service network. At the time, about 3000 corporations were operating some 3500 jet and turboprop airplanes. Those corporations were our target market within the larger general aviation community. We could not question the comparative sizes of our sales force. One hand was all we needed to count our entire sales force, with fingers left over for our support department. However, there were other factors to consider.

The interest of a U.S. corporation to act as our exclusive distributor signaled that we were not the only ones who believed in our future. It also prompted us to examine our situation and to ask ourselves several questions. Should we enter into a business relationship with the distributor or continue to manage our own business? If we retained direct control, what is right and what is wrong with our current approach? How can we improve?

It was finally time to develop a strategy and organization better suited to our new business environment. In 1977 the Avionics Division military business was flourishing, and so we did not necessarily need the civilian market for the foreseeable future. Our concern was what might lie beyond the horizon. We could not anticipate the end of the cold war, the collapse of communism, and a downturn in military spending, but those events were always a possibility. One result of our efforts in the 1970s and 1980s is that today the company has a better balance of military and commercial products and markets. Because we were always under pressure to be profitable, it was harder to take bold new initiatives, especially initiatives that required a sustained investment.

Just one month after the U.S. distributor approached us, we decided our course of action. That decision paved the way to a coherent approach to the design and marketing of our commercial avionics products.

The reasons for our decision and what we did are important to this story, and so I will summarize the principal observations that we made during that month of self-assessment. We would not have succeeded if we had not been completely honest with ourselves.

We must always be conscious of changing business circumstances. Successful companies recognize the need to change, know what to do, and then accept the challenge of change. Sailors who choose to stay the course without taking heed of the weather forecast may end up capsizing. The same standard applies to business.

To understand what we did back in 1977 and why, a few background comments are necessary. For several years military organizations had realized that for many of their operations they could use the less expensive commercial, rather than the specially designed military, equipment. Most of the time military transport aircraft operate in an environment more similar to that of commercial airliners than that of jet fighters. Also, by buying commercial airline avionics, the military organizations could use the specifications that the airlines had written. Then they did not have to waste time and money developing their own voluminous specifications. Our Omega equipment was one such product line that could be used by both commercial and military operators.

That fact led us to think in terms of commercial avionics products not markets, and a class of products that we differentiated by their design, not by their customers.

Elsewhere in the division, a group had developed a set of aircraft engine instruments, which were unique because they had a vertical display and no moving parts. The situation resembled the competition between digital wristwatches and round dial watches. Customer acceptance of our new instrument technology was similarly split.

The U.S. Army ordered hundreds, then thousands, of our instruments for several of its helicopter fleets. The vertical instruments were particularly attractive because they needed less cockpit space than the round dial instruments. Operators of fixed-wing airplanes, however, were more cautious about adopting our idea.

Because private business jet airplanes have small cockpits, we thought that community would be interested in our instruments. If it was, we would have another product for the commercial market.

The last point that I want to make before discussing our November retrospective assessment concerns our approach to the general aviation market. By the end of 1977, we had signed agreements with 22 fixed base operator dealers; 5 were in Canada, and 17 were in the United States. Those companies helped to sell our Omega equipment to private business airplane operators and installed the equipment. Those companies were becoming stakeholders in our business, an attitude that we encouraged.

We held many formal and informal meetings during November and consumed more than one jug of beer at the Klondike Tavern in the evenings while we decided many of our important strategies. Carl Hanes, our solitary general aviation marketing person, provided suggestions about how we should handle the general aviation market. Peter Gasser, Ervin Spinner, the division marketing manager, Lionel Leveille, our manager of product support, and I provided a broader assessment of our situation. We felt that the time had come to reassess our entire approach to the commercial avionics business, not just general aviation.

Carl pointed out that general aviation avionics manufacturers that had what he called uncomplicated avionics used distributors to sell their products. That was the only way they could effectively reach the thousands of individual customers, each of whom bought only one or two systems at a time. Carl's uncomplicated avionics were those that were required by every aircraft and were easy to install. Compasses and radios fell into that group, but not Omega equipment.

The distributors had a network of dealers, many of whom were also fixed base operators. In this situation the avionics manufacturers remain in the background, except when something goes wrong; then they are the ones the customers blame, not the distributor or dealer.

By using distributors, equipment manufacturers avoid having to set up a costly distribution network of their own. However, the manufacturer–distributor relationship has to be a true partnership. The manufacturer's work is not completed once it has processed a distributor's order. Marketing and service must be a team effort. More importantly, manufacturers must remain in direct contact with their customers. Only then can they expect to maintain their competitiveness and to be responsive to their customers' needs. The more complex the product and the more limited the market, the more risky the distributor approach to marketing.

A distributor and a dealer, between the manufacturer and the customer, also increase the price of a product. Our CMA-734 was not only the smallest but also the least expensive Omega Navigation System available. If we used a distributor, the price of our product would have to increase by 20 percent, and we would lose our price advantage.

What concerned us the most about the use of a distributor was that we would be farther away from our customers. If we were to be successful, we knew that we had to remain in direct contact. Only by maintaining contact with our customers could we hope to understand their long-term product needs. That knowledge was an essential ingredient to our discussions because we planned to develop a range of related commercial products, not just more and better Omega equipment.

If some other company, less expert than we were, was going to assume the responsibility for solving customer problems, we could end up with more problems and a bad name. The Omega System was still a new form of navigation, and we expected that we would have to solve many more problems.

An agreement with a distributor also would mean that our fixed base operators would have to work with the distributor and not directly with us. We enjoyed good working relationships with those companies and, in some cases, friendships, as with Tom Kokozinski of Butler Aviation. Canceling those agreements because we had deficiencies in our own organization was no way to reward those companies for their hard work.

Perhaps later, some form of distributor arrangement would benefit all parties involved, but not now. To build our business, we still had to accept direct responsibility for much of the work.

Increasing our customer base and improving our market share were our longer-term goals. At the time of our internal deliberations, our market share in general aviation was less than 10 percent. Global Navigation Systems Incorporated and the Communications Components Corporation together had captured some 80 percent of the market.

With a distributor we might have captured 25 percent of the market during the following two years. However, we determined that we could not maintain such a share unless we took direct control of our marketing.

It was now clear to us that we had to change our work habits and, if necessary, our organizational structure to succeed in the civilian market.

It was time to be honest with ourselves. Our organization, with small business units for each product line, was working well for large military and civilian contracts. With this organization, a customer only had to contact one organization within the company. What was wrong with this organization? Too many customers, and more being secured daily, coupled with severe staff shortages, particularly in marketing and field support, were the problems.

It was impossible for one person to contact all of the 3000 potential general aviation customers. Also, Tom Kelly and his team of support engineers could not handle all the requests for advice and support.

This situation led to two major problems. First, Carl Hanes had to concentrate his general aviation marketing efforts toward our fixed base operator dealers. We were losing contact with our potential customers. Second, as our customer base grew, we were getting more telephone calls for information or help, which was a more serious problem. Carl and Tom handled each other's calls when one was out of town, but often both would be away. Then Peter Gasser, one of the engineers, or I would handle the calls.

Dealing with those calls and other activities for our colleagues was time consuming. Rapidly, we were reaching a point where everyone was spending more time doing someone else's job. The problem was most serious among the engineering staff because those individuals traveled the least. The engineers frequently found themselves investigating an installation problem or a new problem with the Omega signals. They were always willing to help a customer. Also, investigating an obscure installation problem was often more fun and challenging than testing a new circuit or software. The engineers were content to be diverted, but our mainstream engineering activities were falling behind schedule. We could allow that situation to continue.

Peter Gasser summed up his observations in a memorandum. He wrote in part: "Let's try to eliminate some of our basic internal problems before we go looking for unknown new ones [distributor].... My proposal rests on the principle of doing more work with the same staff available (with exceptions of course), by finding ways to improve the working efficiency of the people involved."

That quote shows that we approached our problems by first looking for more efficient working methods. We did not immediately conclude that we needed more people.

Adding people often aggravates the problem they were intended to solve. More people add to the administrative burden of the group. Most managers will attest that they spend considerable time, or should, on personnel matters. Managing more people means having less quality time to counsel and to help subordinates. Building a team and guiding the business should be a manager's priorities.

For some time, I had thought that we could improve the efficiency of our commercial avionics products marketing and product support if we adopted a more centralized approach. We had examples of redundancy, such as when I would go to Brazil one week to sell Omega equipment, and the next week a colleague would go to Brazil on business for the Doppler product line.

One of my often-repeated statements was "'That's not my product' is *not* a marketing answer to a customer's question."

Peter Gasser and I discussed how we should reorganize to better cope with our new situation. Fortunately, Peter and I had similar views. We had become friends, and I do not think that we could have made our reorganization work if we had not trusted each other. More importantly, we were team players. Neither of us was trying to advance his career at the expense of the other. Peter, as product manager, had the most to lose by any move toward centralization.

By the end of November, just one month after we had started our self-assessment, we had the beginnings of a new organization in place. Fortunately, we found a way to retain the advantages of the product management matrix organizational structure that Kieth Glegg had introduced 10 years earlier.

We created a Commercial Avionics Marketing Department, which Peter asked me to manage. We also created a Commercial Avionics Field Engineering Department. Pierre Fournier, our customer training instructor, returned to the Training Department, where he could work more closely with Russell Kelly, our training manager. No one affected by this reorganization changed jobs. It was an attempt to provide a clearer definition of responsibilities. The reorganization also helped to relieve the product manager and project engineers of routine product marketing and support tasks.

This reorganization was the beginning of the creation of a Commercial Avionics Products Group. It occurred, after several more interim steps, in 1982. I consider the formation of the Commercial Avionics Product Group to be the milestone that confirmed the achievement of our long-term goal to set up a viable civilian aviation avionics products business entity. It is remarkable that we achieved that goal within 10 years of entering the civilian market. Most new business ventures take longer to mature.

The reorganization consisted of small changes, but they had a significant impact on our ability to cope with our new market. Often organizations only need refinements to solve problems. Managers usually reserve a major overhaul of an organization for a true crisis or the formation of a new business unit. Our changes, though small, were not without risk. When Kieth Glegg changed the division's organizational structure in 1968, he deliberately gave the product managers complete control over a product line.

By moving toward a more central organization for marketing and product support, we risked diluting the product manager's authority. We could lose some of the real benefits of the organization that had served us so well for 10 years. We were tampering with the culture of the division.

We averted any problems by keeping intact the management of our exclusively military products. This experiment only involved our commercial products and was designed to solve the special problems we were facing with the much larger number of smaller customers.

The commercial avionics marketing personnel remained physically located in the Omega Lab. Our support engineers were given their own area close by. Despite the functional separation of marketing and the physical separation of the product support group, we continued to work as a team. Division management and our culture assured that the product manager was still in control of the product line. However, we had a clearer division of responsibilities.

At the beginning of December, I began putting together the Commercial Avionics Marketing Department. At first, we thought that we would add one more person. That did not happen for another year. Our original plan had marketing physically separated from the engineering laboratory. Although I was a proponent of that idea, I was later glad that a shortage of space prevented us from carrying out our plan. Our proximity to the heart of the product's engineering activity allowed us to keep current with the technical developments. Also, the engineers had the advantage of knowing immediately when we received another order. The frequent feedback helped to maintain morale and a team spirit.

The marketing department had the responsibility for marketing our Omega Systems to all classes of customers. Also, we were tasked with marketing the division's engine instruments to the civilian market. The instrument program retained the responsibility for its military sales and customers. It was a modest start to adjusting our business to serve the needs of our new market.

In 1977 the Omega product line contributed less than 5 percent to the division's annual revenue, but out of more than 15 design projects being undertaken within the division, the Omega product line was the largest consumer of the division's research and development budget. What set the Omega product line apart was that it was the only product line that was capturing a civilian market. The company's management understood the importance of securing a diverse market for continued profitability, and so it was prepared to make that investment to achieve a long-term, balanced, and growth business.

We did not have a corporate strategic planning bureaucracy that was projecting a downturn in the military business. On the contrary, several departments within the company were actively involved in projects that would broaden the scope of our military business. Our emphasis was on manufactured goods, though we were starting to expand into engineering services. We had based our decision to invest in the civilian sector of our industry on logic, not analyses.

In this account I rarely talk about our corporate management. Although it was active in our business, at both the strategic and tactical level, most decisions were the result of recommendations from the line organizations. The company's corporate management encouraged those with the first-hand knowledge to plan, as well as to execute, the business.

To be fair, without our successful military products we would have had difficulty investing in the Omega product line. The Avionics Division had a profitable Doppler Navigation Systems business unit. That was no longer a growth business because the military was using Inertial and Omega Navigation Systems for all operations except those where a Doppler system had a unique advantage. The military engine instruments were becoming a major source of revenue. Our

Defense Communications Division was the Western world's supplier of an army tactical radio system. Those and other products gave us the financial security for new ventures.

However, the company was investing in more than just the Omega product line. For several years we had been making our own specialized electronic components and printed circuit boards. We were investing to convert that capability into its own business unit. The British Telecommunications order that launched our distributed processor telex exchange product provided the stimulus for continued investment in that business. The Avionics Division was investing in several other new technologies and products. The Omega product line was just one of many investments. We did not have a special status when it came to justifying our development and marketing budgets.

The realignment of our commercial avionics organization allowed me to start improving the efficiency of our marketing operations. Peter Gasser, our product manager, retained his authority for all phases of the product's life cycle. He also concentrated on keeping the engineers focused on their assigned tasks. Ray Thibotot, who had been the Omega product line manufacturing manager, became the manager of our newly formed Commercial Avionics Customer Support Engineering Department.

The popular Jean Duplantie was promoted to be our manufacturing manager. His skills, his passion for his job, and his understanding that it takes a team to attain objectives were critical ingredients to our success. He and his assistant, Eddy Boudreau, worked wonders during the years when we had more than 40 versions of our Omega products in production at the same time.

Russell Kelly and Pierre Fournier, now working more closely together, could concentrate on implementing a customer training program. Hans Koller, the manager of our Repair and Overhaul Department, worked to set up a worldwide, around-the-clock repair service organization for our commercial customers. Hans' dedication to product support was infectious. Where a customer was concerned, nothing was too much trouble for his team of technicians and administrators. George Stinson, head of our Technical Publications Department, guided his writers in the completion of the Omega technical and operating manuals. Because Omega signal problems continued to plague us, Robert Baillie was given the overall responsibility for conducting the special analyses that had become so time consuming for several of our engineers.

By separating, at least functionally, the engineering, support, and marketing activities, we all could concentrate on our own responsibilities. An organization's culture, supported and encouraged by corporate management, can help it to withstand many changes. Our firmly held belief that we could best serve our customers, and ourselves, by maintaining smaller business units also helped us to maintain a cohesive product team.

Our plant was filled to capacity, and our location in a residential area prevented expansion beyond the city block that we occupied. Therefore, when we moved people, it was not far from the product's core, the engineering laboratory. Because of the space limitations, most of us stayed close to the engineering team, which helped us to communicate on a regular basis.

"Management by wandering around" is a popular concept. However, it requires considerable self-discipline. A ringing telephone can divert even those with the best intentions as they are about to step out of their offices for an hour of wandering around. Most people will regularly communicate with those who are within 200 feet of their work-space. Communication deteriorates rapidly after 200 feet. If stairs, a highway, or an ocean separate them, the communication problem becomes much worse. Although physical separation remains a problem for effective communication, the proper use of today's teleconferencing technology, augmented with easily produced visual aids, can help many operations.

My primary concern with our marketing was that individual marketers were only responsible for one product. Their performance appraisals were done by their product manager. The product managers consulted with Ervin Spinner, the division's marketing manager, but the emphasis was on how well the individuals had succeeded with the products to which they were assigned.

I started to promote the idea of multiproduct marketing, but within the context of our product management organization. My concept of multiproduct marketing was that we should continue to specialize in one product line but have a basic knowledge of our other products, enough knowledge to be able to identify a potential market. Then, when Omega marketing people went to another country, they could assess the market potential for our other products with some degree of accuracy. Customers only want to talk to experts; however, there is rarely any harm done if the first contact is at a more superficial level. Customers will excuse marketing people who admit that they do not have adequate answers for the questions, provided they offer to get answers or access to an expert who has the answers. Then, to be successful, the follow-up must be swift.

We found a way to maintain our product line orientation but improve the efficiency of our marketing by introducing a form of multiproduct marketing. I do not favor centralized multiproduct marketing groups. The people in them are usually too far from the realities of the products and lack the detailed knowledge that customers demand. Our concept proved to be an effective compromise.

One of my first tasks, when we created the Commercial Avionics Marketing Department, was to talk to Russell Kelly, our customer training manager. As I expected, Russell was enthusiastic about my suggestion that he and Pierre Fournier develop special product courses for our marketing staff. Within two weeks Russell and Pierre had courses ready, and the four of us in commercial avionics marketing went back to school. Soon after, we convinced the application and marketing engineers on the other avionics products to start attending the courses. It was not long before engineers, contracts staff, and others requested course time. We had made a small, but significant, step toward a form of multiproduct marketing that suited our organization and was effective in the field.

This new-found knowledge also helped to promote better communication within the product lines. The product management organization, with all the product disciplines within four walls, had many advantages. One of its disadvantages, though, was that it made us rather insular. Now the walls, figuratively speaking, were coming down.

We began to improve our sharing of market information and procedures that individual product lines had established. For example, whenever we prepared a proposal, we had to seek management approval for, among other items, the terms and conditions we wanted to offer. Often we were responding to a set of terms and conditions that our prospective customer had specified. However, half the time, especially in the commercial market, our customer told us to propose our own terms and conditions.

In the process of responding to many such bids in the Omega program, we developed a standard preapproved general terms agreement. This meant we did not have to go to corporate management every time we wanted to use those commercial terms. It was not long before the military product managers were using a variant of the general terms agreement developed by the Omega product team, thereby saving themselves much time and effort.

There were many examples of this grass roots initiative to improve our internal communications and support for each other. The company was nurtured by a management that encouraged ideas and only very rarely stood in the way of change, especially changes recommended by line staff. It was, and remains, a dynamic working environment and one in which it was easy to be a team player.

I only remember one case when a manager resisted a recommendation for an efficiency improvement. Although we had developed a standard general terms agreement for our commercial customers, it had to be customized for each proposal. That meant that key management personnel at least had to review the changes. By 1981 we had been through that procedure hundreds of times and had spent thousands of hours in the approval cycle. We knew enough about the business to put together a terms and conditions handbook, which would include all the clauses and their variants. I realized that if we could get such a set of clauses preapproved, we could streamline our approval process.

I spent many evenings at the family dining room table putting together such a terms and conditions handbook. However, one manager resisted the concept. I now believe that I was up against a manager who feared the loss of authority. To his credit he spent many hours at home reviewing our proposals. That he was spending a lot of time on that poor return-on-time-invested activity did not seem to matter. He rarely found a significant problem. It was not until Tony Sayegh became the division's contracts manager that the terms and conditions handbook was approved. It saved us countless hours of proposal writing and approval effort.

Until we created the Commercial Avionics Marketing Department, the marketing budget had been a line item in the product's engineering budget. It is difficult to hold people responsible for activities for which they do not also have the budget authority. With the formation of the marketing group, we separated the marketing and engineering budgets. I further divided the marketing funds and made all the marketing people responsible for their own travel budgets.

Separating the budgets also gave us an opportunity to streamline some of our other marketing activities. We started to better coordinate our approach to advertising, trade shows, brochures, and the variety of other marketing support functions. Though our magazine advertisements promoted a particular product, we coordinated their schedule for maximum company exposure. We no longer placed three advertisements for different products in one issue of a trade journal but instead placed advertisements in three successive issues. Advertising agencies say that people read only the header when they first see an advertisement. In the next issue, they will read the first and second lines. It takes four consecutive placements for most advertisements to be read in their entirety.

Within a few years we set up a Marketing Services Department for the division, which provided a centralized service for activities that the individual programs were carrying out and often duplicating. The idea for that department and other efficiency ideas only became apparent when we began to succeed in the commercial aviation market with its multitude of smaller customers.

We were always careful to maintain our product management organization, and we only centralized functions to better serve the product manager. We were anxious to maintain the team and entrepreneurial spirit that our approach promoted. We also knew that we had to stay alert to changing market conditions and continuously fine-tune our organization to meet our new market challenges.

Those of us in the newly created marketing group considered that we had an obligation to bring forward new product ideas from our discussions in the marketplace. We took that obligation seriously, although our first priority was to sell our existing products. Without a broader commercial product base, we could easily become a one product line organization. Then we would lose the advantage of our growing market base. We were in the business for long-term revenues. Success is long-term business with manageable growth. Companies achieve that growth with a long-term commitment and the willingness to work hard to win market share.

It was time to organize ourselves into a cohesive marketing group with a strategy and marketing plans. When we had been responsible for one product line, the world was our territory. Now that we were four people, we could consider splitting up the world into more manageable territories.

During the preceding five years, I had visited more than 30 countries, most more than twice. I had traveled in North and South America, Western and Eastern Europe, the Middle and Far East, and the South Pacific. I was beginning to understand the various markets and their interrelationships.

I split the world into three major territories: North America, South America combined with Asia and the South Pacific countries, and Europe combined with Africa and the Middle East. That territorial division served us well for many years. I did not claim any intellectual wisdom, only an understanding of the markets. However, that territorial division is close to the evolving economic unions within Europe, the Pacific, and the Americas. Today, for marketing purposes, I would combine South America with North America, not with Asia.

We were only four people, and so our responsibilities had to be further defined. Within those territories, we assigned ourselves primary and secondary product and market responsibilities. Markets included the world's general aviation companies, airlines, military organizations, and airframe manufacturers. We also covered intercompany arrangements with Marconi Avionics in England and Marconi Italiana.

An important feature of our approach to marketing was that the marketers stayed with their customers. When we came back from a trip, we did not hand the paperwork and follow-on activities to someone else. Most of the time there was no one to pass that work on to anyway. We often wrote the technical and commercial proposals, coordinated evaluations, and arranged for field support and training. Even after a sale we acted as the primary point of contact for many of our customers, particularly our overseas customers. However, Gilbert Champagne provided our marketing group with invaluable, and always cheerful, contract administration support.

We prepared brochures and wrote trade journal advertisements. During the early period of the Omega equipment launch into the commercial market, when we were all learning, we helped to design the pilot's operating guidebook. We gave seminars to the regulatory agencies and helped with the product flight certification. It was not a time to say "that is not my job."

The varied in-house activities that our marketing personnel had to perform put a strain on our travel time. I calculated that anyone who was spending more than 30 percent of work time on travel was reaching overload. Sometimes we traveled 50 percent of our time. Somehow, though, we also completed the other tasks.

By the end of 1977, we clearly were a lean organization. The division of responsibilities appeared complex, but it worked because we refused to acknowledge that the job could not be done. We were young and eager to triumph over our competition. We also could rely on support from the engineers and many others within the organization. We were still a team, though we had a better and more efficient organization.

In the summer of 1978, our marketing organization expanded by 25 percent with the addition of Tony Deveau. He had been a technician for our engine instruments group and was eager to join the marketing team. Tony came to us as a marketing trainee, but it was not long before he became productive in the field.

In September John Simons promoted Peter Gasser to group manager. Peter was responsible for several product lines, including the Omega product line.

When Peter became group manager, he recommended that I take over as the Omega product manager. That came as a surprise, but one that I welcomed. With Peter as my group manager, we could continue with the partnership that had worked so well in the past; however, in late 1978, we had a few more responsibilities. I continued to lead the commercial marketing effort.

As product manager, I was fortunate to have competent project engineers working for me. I could rely on their judgment and their skills in leading the day-to-day activities of the engineering team.

Jean-Claude Lanoue, our seasoned software and hardware engineer, who had more innovative ideas in his head than we could cope with, was leading new Omega design activities. Robert Baillie led the sustaining engineering work for our Omega products that were in production. Those were the CMA-719, our first-generation system, the CMA-734 for general aviation, and the CMA-740, our first airline Omega Navigation System. Robert was helped by David Bailey, who had been with us for five years, since his graduation. An accomplished engineer, David was destined for early promotion within the company which he rightly deserved. As if to confirm his potential, David attended night school to earn a master's degree in business administration.

When I took over as product manager, I had to better understand all the design activities that we had under way. As usual, when I wanted to understand something, I wrote it down, in this case as activity flowcharts on large posters. I was anxious to make sure that all the engineers understood what was going on in the Omega Lab and in what direction we were going, and so I festooned the walls with my charts and held staff stand-up meetings by the charts. Many years later, I have found that morning stand-ups are a regular feature of Washington, DC, life. I have also found that those regular staff meetings are an effective communication tool.

By 1978 our engineering team had completed the designs of our CMA-734 general aviation and CMA-740 airline systems. They also were working on a variant of the CMA-740 to meet the needs of our new airline and military customers. We had other design work under way that would assure a larger market for our products. So far, operators had bought Omega equipment to replace out-of-date navigation systems. We knew that this market would soon be completely satisfied and that to sustain our business we had to offer an Omega Navigation System that was useful to other operators. How we took the initiative to expand our market is a story for another chapter.

By the end of 1978 we were maturing as an organization, one fully committed to commercial aviation. That year our general aviation sales doubled. We sold more than 150 systems into that sector of the market. We expanded our fixed base operator dealer network from 22 to 45 companies. Two were in Europe, and the rest were in North America. Eighty percent of our general aviation sales had been in the North American market, which has the largest concentration of corporate business jet aircraft operators. That market also has the most fierce competition.

Six of the leading general aviation airframe manufacturers had ordered our systems, and we were making good progress with Marcel-Dassault Aviation in France. We had solved the problems associated with installing Omega Systems in a variety of different aircraft, including helicopters.

The airline community now considered us to be a bona fide airline avionics supplier. We had proven our ability to meet the complex product and support requirements of this industry. Although we had not been particularly successful selling to the U.S. domestic airlines, we were succeeding elsewhere in the world. We had contracts with some 15 commercial carriers. The major carriers that were our customers included Pan Am, Varig, China Airlines, Air New Zealand, British West Indian Airways, and Monarch Airlines in England. Final negotiations with Japan Airlines were under way as 1978 came to a close, and several Boeing customers had bought our system with their new B-737 and B-727 airplanes.

We were confident of our ability to secure a firm market base in South America and throughout the Pacific. Although the North American market continued to be a priority, we considered Europe to be our next major challenge.

1978 ended with the disappointment of not winning Britain's Royal Air Force contract. However, elsewhere in the military markets we had reason for optimism. We had secured an order from the Chilean Air Force, but only after several visits and a flight test. Both the McDonnell Douglas Aircraft Corporation and Beech Aircraft had selected our system for special purpose U.S. Navy aircraft, and Lockheed-Marietta had selected our system for C-130 aircraft. We were actively marketing to 10 other military organizations in 6 countries.

During 1977 and 1978, we matured as an organization, and we had positioned ourselves to secure long-term business, particularly in the commercial sector.

14

Commitment to a Market

*Success is not the result of spontaneous combustion.
You must set yourself on fire.*
—Reggie Leach

To succeed a company and its people must commit themselves to their customers and their business. Commitment means enthusiasm. It means working to deadlines and budgets, doing your best work, making an extra effort, being responsive to customers, and much more.

The Canadian Marconi Omega team members repeatedly, and over many years, showed their dedication to our common goals. They tackled technical problems and overcame the disappointment that followed a competitor's success. Fortunately, even before we had our first big success in the airline market, we had the support of the company's management. Without that support at all levels of the company, we would not have succeeded.

Although a dedicated team and an obsession to win are important constituents to success, a commitment to the market is essential. Sustained success is the reward for a continuous investment in, and attention to, customers.

Believing that the only continuing investment necessary is a good marketing team will not bring success. Marketing professionals are the first to admit that they cannot sell a product that does not meet a customer's needs. Customers are knowledgeable; they know what is possible, what they want, and what they can afford. Marketing people who believe they can sell anything to anybody are not professionals, and nobody should want an amateur.

Justifying an investment, like commitment, can take many forms. Some companies have only one measure, a rigid calculation of the financial return on investment. If the expectations are unrealistically high, that will hamper a company's ability to succeed. In the beginning, we justified our investment in market, not solely financial, terms. Our investment allowed us to enter the airline market. Our commitment helped us to build a market. Then we reaped the benefits of a good financial return on our investment.

Our first customers, those who had to replace their LORAN-A navigation systems, were initially influenced more by price than performance. Omega Navigation Systems were much less expensive than the alternate Inertial Navigation Systems. Although some airlines, like Pan American, had flight tested systems enough to know that Omega was a trustworthy form of navigation. Most other airlines remained skeptical. That is, until finally we had seven Omega stations on the air, and the pioneering airlines had proved the potential of the system. This all came together during the autumn of 1976.

In October of 1976, the Airlines Electronic Engineering Committee set up a subcommittee to write the specification for an airborne Omega Navigation System that would satisfy the needs of the entire airline community. To stay in the market we had to face the reality of launching the development of a new system, a system that we knew would replace our as yet unsold CMA-740 system.

The result of the committee's work was a specification called the ARINC-599 Characteristic. It described an Omega System that could connect with several aircraft instruments, the aircraft's autopilot system, and other navigation systems. We called our version of that equipment the CMA-771. It was a direct variant of the CMA-740. In that way we could better serve our CMA-740 customers because we could offer them upgrades.

By making the two systems compatible, we ensured that our existing customers would not have to buy new equipment as their requirements changed. Also, they might have requested bids from other suppliers, and we may not have won the follow-on contract. Our approach was self-serving, although at the time we were thinking more of our customers and of reducing our production costs than of developing a wise business strategy. Once again we relied on our intuition and did what was right rather than follow an established strategy.

When we saw the benefits of that approach, we maintained it for many years. As a result, our customer base grew, and we did not lose customers to other suppliers. We remained loyal to our customers, and they remained loyal.

Eventually, advances in electronics technology and our customers' desire for smaller, lighter-weight, and lower-powered products prevented us from maintaining complete upgrade compatibility. Some 17 years after we first introduced our CMA-740 Omega System, Canadian Marconi was still selling and upgrading equipment that had its origins with the first system, although the similarity would not be easy to recognize because the later systems had capabilities and features that were unthinkable in 1976.

Most manufacturing companies would benefit from this type of commitment to their market. Black and Decker, a power tool manufacturer, learned this lesson a long time ago. After you buy its basic powered hand drill, you can, with small incremental investments, convert the drill into a temporary bench saw, sander, and several other tools. Once hobbyists realized the utility of those tools, they usually buy tools dedicated to the task. If they started with Black and Decker, they will usually buy a Black and Decker specialty tool. The company that commits itself to the real needs of its customers will keep them as customers.

The Varig Story

The Varig story illustrates that companies that are committed to their customers and their products can succeed against severe competition. The story started in 1974, when the airline flight tested our first Omega Navigation System. Varig held off making a purchasing decision due to the immaturity of the Omega ground network and evidence that several manufacturers were designing less expensive Omega equipment.

Despite that delay we continued to visit Varig. Our marketing took the form of keeping Varig up to date on progress with the Omega ground stations and our new airborne equipment design work. We did not try sell to Varig equipment that it did not want or did not need at the time. We recognized that we had to be patient.

I could have subtitled this story "Or How We Kept a Promise Made in a New York Taxi Cab." That was when I had reluctantly promised Assis Arantes of Varig an accuracy of 4 nautical miles 95 percent of the time, anywhere and at anytime on Varig's routes with a system that still had many unknowns. Later, I could have claimed that I meant 95 nautical miles 4 percent of the time and that a particularly large New York pothole had caused me to transpose the figures.

Varig started flight testing our second-generation Omega System, the CMA-740, in May of 1977, one month before we received our product launch order from Pan Am. It tested our equipment for more than 12 months. Varig flew our equipment, and that of our competitors, to the east and west coasts of the United States, and to Europe, South Africa, Chile, Argentina, Mexico, and Japan.

Originally, Varig had intended to complete its flight tests in October of 1977. It extended the tests by nine months because of problems with all the competing systems. The more often we flew on Varig's routes, especially those across the South Atlantic to Portugal, the more we realized that the Omega System had a serious flaw. The Earth's magnetic field and the sun's effect on the ionosphere were disturbing the Omega signals as they crossed the equator.

The first few months of the test period were a big disappointment. No system was working properly, and most had navigation errors as high as 100 nautical miles when they crossed the equator at sunrise. Varig started to think about buying Inertial Navigation Systems for its B-707 fleet, even though they were four times the cost of Omega equipment. The airline's newer DC-10 aircraft used the Inertial Navigation Systems, and the pilots were pleased with their performance.

We could not afford to fail with a potential customer, particularly one as strategically important as Varig. Airlines are in the business of transporting people and cargo. They are not in the business of providing a vehicle to help design new equipment. Even so, Varig showed a considerable commitment to the Omega System.

With Varig's help we set up a flight test program to investigate the Omega signal problem. Like Pan Am, Varig was prepared to assume the role of pioneer to help the industry. That generous and professional approach can be attributed to Assis Arantes. He faced much opposition to continuing from within Varig, but he knew that if we succeeded, he would save his company a considerable amount of money.

We assigned two engineers full time to the task of analyzing the flight data and investigating modifications to the equipment. Three of our field engineers took turns flying on the Varig B-707.

In the meantime, the FAA had designed special aircraft recorders to collect Omega signal data to support its certification activities. We borrowed a recorder and installed it on the Varig airplane, while Robert Baillie, our Omega signal propagation expert, designed an elaborate set of laboratory equipment, to analyze the digitally recorded information.

That analytical facility became one of the most comprehensive of its kind in the world. It allowed us to build an extensive database of Omega signal coverage and propagation characteristics and over time to find the causes of erratic Omega signal behavior.

Our success was helped by our willingness to recognize, take responsibility for, and search for solutions to problems, even problems not of our making. We did not hide information from Varig. During the test period, I telephoned Assis Arantes every week to tell him exactly what we had found, what information we still needed, and whether or not we were making progress in finding a solution. It was a cooperative effort.

We found solutions to the problems, and in May 1978 Varig selected the Canadian Marconi CMA-740 Omega equipment. Once again our competition had been five major U.S. avionics companies.

In September of the same year, I presented a paper, at the request of Varig's director of procurement, Arthur Mueller, at a meeting attended by representatives of most of the airlines in South America. The paper, titled "Omega Navigation in South America," was, as requested, nonpartisan and directed toward the needs of procurement personnel. In it I described the Omega System, the different airborne equipment, and the airline specifications. That meeting was another occasion when we put competitive considerations aside to help the industry.

One year after Varig installed our equipment in the B-707s, it changed, by one hour, its flight schedule between Rio de Janeiro and Portugal. That change put the aircraft over the equator at a different time during sunrise, which was enough to uncover a new Omega signal problem and set us off on another nine months of investigative work. Our commitment to our customers was such that, regardless of cost, we addressed and solved every problem that came up.

In September 1979 Assis Arantes and I presented a joint paper at the fourth annual meeting of the International Omega Association. We titled our paper "The First Twelve Months—The Varig Brazilian Airlines Omega Program." In the paper I wrote the following:

> "This paper is an account of a remarkable achievement. In the space of twelve months Varig Brazilian Airlines has selected an Omega System, brought it to operational maturity [in B-707s], and achieved maintenance independence.... The airline has also, in this time, engineered a DC-10 Omega installation.... The facts speak for themselves, and this achievement demonstrates the high calibre of professionalism that we have found to abound within Varig.... The [Varig's] technical, administrative, and operational assistance at all levels has on many occasions been beyond that which we might have reasonably expected. The established working relationship between the two companies has helped to make *The First Twelve Months* a success."

Assis wrote the following:

> "The early results of the evaluation with most systems were disappointing.... By March 1978, the system choice was becoming easier.... The selection of the [Canadian Marconi] CMA-740 was based primarily on the in-flight performance the system exhibited during the evaluation period.... The manufacturers' willingness, during the preceding nine month period, to work closely and openly with Varig on mutually beneficial programs was also considered [during the selection process]."

Both companies had committed themselves to the project, and both companies had benefited.

Boeing began fitting our equipment into its B-737 airplanes in 1978. Any problem that prevents an airframe manufacturer from delivering an airplane on the scheduled day unleashes a frenetic amount of work to solve the problem. The most senior managers in the company become involved, and they subject any supplier that is responsible for the problem to a lot of pressure. When millions of dollars are at stake, the loss of income, the interest on that income, and even the cost of extending the insurance for the airplane become significant. Suppliers to airframe manufacturers live in fear of such events.

Our first B-737 customer was the Liberian government. While Boeing was training the Liberian flight crews, a problem recurred that we thought we had solved. When the pilots connected the Omega equipment to the autopilot, the aircraft oscillated in a roll.

What followed was a period of long work days and canceled Christmas holidays. We sent engineers to Boeing, and they sent engineers to our laboratory in Montreal. For weeks the problem eluded some of the best systems engineers in the business, but eventually they noticed that the autopilot oscillations were not as random as they appeared. They occurred only on certain aircraft compass headings.

Finally, the Boeing compass system experts tracked the problem to the calibration of the compass. On their own, the compass system, the autopilot, and the Omega equipment were working correctly. The problem was that, when working together, one system affected the other. A small change to the calibration of the compass solved the problem.

We had much other priority work, and we could have delayed helping Boeing by asking for more information—but that was not our style. It is also no way to treat a customer or to succeed. We tackled all problems like that one with equal commitment and intensity.

Our product management matrix organization allowed us to judge quickly and to change our priorities without following a bureaucratic process.

Sometimes a company can improve its reputation through an emergency. The problem that we experienced with our first installation at Boeing gave us an opportunity to show Boeing that we would spare no effort to solve a problem; that is commitment. The incident gave Boeing confidence in our ability, and we enjoyed many years of business with Boeing.

We learned from our customers and our competitors and by failing to win a few contracts that to remain competitive we had to provide more than just equipment and its support. When an airline first installs a piece of equipment, it has to prepare installation drawings and other documents. Those documents have to be reviewed by the national aviation regulatory agency before the agency will issue a certificate approving the airline's use of the equipment. The preparation of those drawings and associated documents takes time and money. They are also transferable to other airlines. Unlike our competitors, we had not fully understood that process, but we learned fast. We started to buy the rights to the installation drawings that our own customers had prepared, so that we could offer them to others.

Litton also offered to make the aircraft wiring harnesses for its equipment. The company had more experience in the airline market than we did in the late 1970s, and it knew that many airlines welcomed that service. Most customers made their own harnesses, but some, especially those that needed only one for a system flight test, were happy to have the equipment manufacturer supply the harness. We learned about aircraft harnesses and the regulations governing their manufacture. Then we offered that service to our customers.

In the autumn of 1978 we secured a contract to provide an initial quantity of 34 Omega Systems to the Chilean Air Force. The Chilean Air Force became our first customer for aircraft harnesses. We had expected to build one or two a year, not 30 in one batch, but once again our manufacturing department rose to the challenge. We found a place in the building, initially a corridor, where the staff could lay out batches of cable 100 feet long. It never occurred to us to subcontract the work. This was our initiative, and the team members wanted to do it themselves.

Pan Am, like the first airline operators of Omega equipment, planned to use it with the existing Doppler Navigation Systems. To make the operation of the two systems compatible, Pan Am asked us to change the software and the engraving on the system's control and display unit. That meant we had to create a new variant of the CMA-740, with its own part number and documentation.

Although we had only 4 variants of our CMA-740 system due to specific customer requirements, we ended up with 40 different variations of our next airline Omega System, the CMA-771.

The proliferation in variations in the product was not due to a lack of flexibility in the design. It was because we were prepared to give our customers exactly what they wanted, if technically possible.

Some of our customers were starting to install navigation management systems. Those systems had a sophisticated computer and the ability to accept information from the many long- and short-range navigation systems that are installed on aircraft. The navigation management systems also had a unit that could control and display the information from those other navigation systems. That led us to plan, in 1978, for a version of our small Omega System, the CMA-734, that would work with a navigation management system. Our idea was to offer an Omega sensor that would provide information to, and be controlled by, a navigation management system. We were listening to the market and responding to its particular needs.

For many years we had been educating the world about the Omega System, as often as we were trying to sell our own product. As pioneers we knew we had to play the role of teacher, not just salesman.

Early in 1977 the FAA asked me if I would give a two-day seminar to its flight inspectors. It did not want a sales talk. It wanted a nonpartisan description of the Omega System. It wanted to hear about the system's idiosyncracies, its problems, and the solutions that we had found. Also, the FAA wanted to understand the differences between various receiver designs adopted by the Omega equipment manufacturers.

Pierre Fournier, our senior Omega System instructor, agreed to help. To save the flight inspectors the trouble of writing copious notes during the seminar, I wrote a booklet titled *Omega Navigation*. The booklet became a useful reference for many people who wanted to understand the Omega System and its application in the airborne environment. The booklet was another of our contributions to the industry. Over the years we printed and distributed several thousand copies of the booklet.

Hollingsead International, a small company in California, is a classic example of a company with commitment. The company designs and produces racks that hold individual avionics boxes in aircraft. Robert Hollingsead, founder, owner, and president of the company, actively participates in all the major airline industry meetings, particularly those, like the Airlines Electronic Engineering Committee, that are working on the specifications for new equipment.

Hollingsead International regularly issues technical newsletters to tell the industry about new concepts in avionics rack design. The layman may think that an avionics rack is a trivial piece of low-technology equipment. A properly designed rack, though, can significantly improve the reliability of an avionics installation in an aircraft. Robert Hollingsead frequently shares his ideas in open forums to help the industry. I am sure that the company has benefited from his willingness to be a full participant in the industry and to not keep secrets in the hope of gaining a competitive advantage; that is commitment.

Most companies will profess a commitment. However, those with centralized functions, and the cumbersome processes that centralization can produce, may unintentionally stifle the practical application of a timely commitment. Decentralized organizations, with strong corporate support, will most often breed commitment and a continuous search for excellence by all in the organization. Employees are motivated to succeed if they are given authority and responsibility and can measurably contribute to the business.

I have only described examples of the technical commitments that a company must make to its market to succeed. The types of commitments depend upon the industry. Most are obvious to those who listen and learn from their customers, competitors, and government agencies and who commit themselves to their customers and treat their business relationships as partnerships.

As this book will continue to show, our commitment was to producing products that met the needs of our customers. We protected our customers with a comprehensive set of contractually binding responsibilities. We put in place a product support infrastructure that ensured that our customers could rely on our support at any time. We supported our industry's associations and its regulatory agencies around the world in efforts to produce equipment specifications and regulations that best served our customers and the traveling public. We maintained cordial relationships with our competitors and our competitors' customers; we learned from them. Our corporate management supported our commitment, and those of us on the Omega team were committed to our company.

The commitment must be complete and cover every facet of the business. It also must come from every individual within the company.

15

Taking the Initiative to Expand the Market

People are always blaming their circumstances for what they are. I don't believe in circumstances. The people who get on in this world are the people who get up and look for the circumstances they want, and, if they can't find them, make them.
—George Bernard Shaw

For us, launching a product and then transitioning into and building a civilian avionics business was a five-stage process.

During stage 1, we developed our first airborne Omega navigation equipment. Also, we had to convince the world that this new form of navigation was viable.

During stage 2, we developed an Omega product line better suited, than our first system, to the needs of the civilian aircraft markets. We discovered many differences between military and civilian business practices, and so we listened and learned about the commercial aircraft operators' expectations of both us and our products. When we knew what to do, we put in place the business infrastructure necessary to sell to and support the civilian market. All the time we promoted the company's capabilities, its commitment, and our new product. Stage 2 ended when we secured launch customers, among them Pan Am.

With credible launch customers and a suitable business operations environment in place, we were ready to embark on stage 3. We focused on building as broad a customer base as possible. It took us three years to complete stage 3. The end of stage 3 overlapped with the start of stage 4.

By 1979 we and our market were ready for stage 4, the period when a company with a new product or market must find ways to expand the market. There comes a time when the initial sales stop growing or decline. Successful companies anticipate that plateau and prepare themselves for stage 4. If they are successful, they have both a base market and a fringe market. As the customer base grows, what was originally a fringe market may begin to dominate. However, it becomes available because the company has designed new product features or derivative products that satisfy the requirements of a new class of customers.

With an expanding business base, a successful company is ready for stage 5 in the business development cycle when the company can focus on, and reap the benefits of, a long-term sustained and growth business. Product development continues, and so does marketing and the search for new responsive strategies. However, if a company has reached stage 5 it can, with diligence, sustain its business. Also, the company is now ready to expand with new products.

Many companies fail to recognize or succeed with stage 4 and their business declines. Canadian Marconi succeeded.

It took us 10 years to reach stage 4 in the commercial market. The delays in the construction of the Omega ground stations had slowed market acceptance of this new form of navigation. Regardless, we also learned that it takes time to change from being a supplier only to the military to being a supplier to the civilian market as well.

We were well prepared as we completed stage 3 and entered stage 4, the market expansion phase of the business cycle. It was a good time to take stock of our situation.

Our Situation, 1976–78

We had secured the largest, and most prestigious, airline contract with Pan Am. TWA had awarded Bendix a contract for Omega equipment, so we thought that Bendix would remain a formidable competitor. We were surprised when, in 1978, it began to drop out of the competition.

Each contract, no matter its size, had been and continued to be a head-to-head battle among several competing companies. It was a difficult market, not one where a company could achieve a quick return on its investment.

Dynell, which had won the U.S. Air Force Omega competition and business with KLM, Aeroflot, and other airlines, had orders for 800 systems. It led in quantity of systems sold, but we led in numbers of customers. Within a few years, and despite it early success, Dynell gradually withdrew from the Omega market.

As Dynell became less of a threat, Litton emerged as a strong competitor. By the end of 1978 Litton had orders with Eastern Airlines, American Airlines, the British Royal Air Force, and others, and we viewed the company as our biggest obstacle to achieving a worldwide market. One of our earliest triumphs over Litton was Lockheed's selection of the Canadian Marconi Omega System for installation in most of its C-130 aircraft.

Tracor never lost confidence, and it continued to be an aggressive competitor, with its share of successes. We remained wary of Crouzet, an electronics company in France and the only company outside of North America to design an airborne Omega Navigation System.

Global Navigation Systems Incorporated continued to dominate the business aviation Omega market, and the Communications Components Corporation was also a serious competitor. Despite that competition, we managed to make many sales to that segment of the civilian market.

Two features that were common to our three Omega Navigation Systems initially gave us an advantage over our competitors. We offered an optional receiver that allowed the system to navigate by also using the signals from the U.S. Navy communication stations. We also offered several types of antennas because the size and shape of the industry standard Omega antenna was not suitable for all aircraft.

By the end of 1978, the FAA had approved our Omega equipment for use throughout the world and as the primary means of navigation.

We had identified 94 airlines in 51 countries as potential users of the Omega System. We had visited most of those airlines at least once. Most countries have only a few airlines, and some only one airline, and so we had to travel throughout the world. Half of the airlines had expressed a definite interest in our product. In addition to those airlines, we were pursuing six major commercial airframe manufacturers in three countries.

While we pursued the civilian market, we maintained contact with 27 military agencies in as many countries. We also visited the leading military transport and helicopter manufacturers. Fortunately, there were airlines to visit in the same countries.

The four of us on the Omega Systems marketing team—Carl Hanes, Henry Schlachta, Tony Deveau, and me—were scarcely enough to cover 50-plus countries and more than 130 potential customers, but somehow we did. We learned to plan our trips carefully to ensure an efficient use of our limited marketing funds and resources.

Our marketing was creating the usual requests for the loan of equipment for flight test purposes. Toward the end of 1978 18 potential customers were conducting flight tests in India, Europe, South America, North America, Japan, and the Philippines. For each of those tests we had to provide onsite installation help, crew training, and general support during the flight test program. We were stretching our Field Engineering and Training Departments to the limit of their resources.

One of our big hopes was United Airlines, which had finally scheduled a flight test program for early in the following year, 1979. In the meantime, the Canadian forces were buying more of our first-generation equipment, the CMA-719, for their new maritime patrol aircraft. More Boeing customers were accepting the option of our standard airline system, the CMA-771, for their new B-737 and B-727 airplanes. Other airlines, such as Braathens SAFE of

Norway, had their airplanes wired for an Omega Navigation System but did not immediately buy the equipment, at least, not until we followed through with our own marketing.

Pacific region countries decided to follow the U.S. lead and shut down the Pacific LORAN navigation service. That shutdown gave us a market opportunity in the region. China Airlines was the first Pacific airline to install Omega Navigation Systems, and it selected the same version of our equipment as Pan Am.

In expectation of Pacific market opportunities, I had looked for a company that could provide our future customers with a local support service. Without such a service, we would not have a market; we had to take the initiative.

Instead of hiring an independent support firm, we involved Air New Zealand and Japan Airlines, both customers by then, in an arrangement by which we paid them to repair equipment used by our other customers. We had similar arrangements with several of our major airline customers throughout the world. The program worked very well and was just one more example of how we took the initiative with our customers to our mutual benefit.

Cathay Pacific, operating out of Hong Kong, tested our system during 1978, and Bristow Helicopters, Britain's largest offshore helicopter operator, tested our system over the Indonesian offshore oil fields.

The contracts with Varig and the Brazilian Air Force gave us leverage for other opportunities in South America. With the help of our local representatives, we focused on Chile, Argentina, Peru, and Venezuela.

I attribute much of our success to our worldwide network of part-time commissioned representatives. Their presence was essential, as was our diligence in selecting the most effective individuals and firms. Our most successful representatives were those we involved during the entire precontract and postcontract phases of our marketing effort. Some companies believe that all they have to do is hire a local firm or individual and the business will unfold. When it does not, they blame the representative. Instead, they should be blaming themselves. Most likely they did not involve or listen to their overseas representative until it was too late.

We also worked with the Canadian government's commercial counsellors located at our embassies and consulates throughout the world. They helped us to select our local representatives and to understand the business practices of a country.

Closer to home, we had an evaluation under way with AeroMexico. That contract was one we did not win. AeroMexico became one of Tracor's early customers.

Our alliances with Marconi Avionics in England and Marconi Italiana continued to mature and produce results. We provided Omega equipment for a British-made aircraft test in India, and we were pursuing several opportunities in Italy.

One of our most important European prospects by 1978 was Air France. We knew that a contract with Air France would be as strategically important as our Pan Am, Varig, and Japan Airlines contracts.

As so often happens, the situation changed, and Air France delayed purchase of new airplanes. Throughout, we remained convinced that Air France was worth pursuing. We supported flight tests by the airline, performed analyses of its route structure, and made cost of ownership studies, all to show Air France that Omega equipment would help to make its flight operations more efficient. We kept France's civilian aviation regulatory agency informed of our activities with Air France as a safeguard against the day when Air France might approach the agency to certify our Omega equipment. After four years of marketing to Air France, it equipped its B-727 fleet with Canadian Marconi Omega Navigation Systems.

The Middle East was one market we did not have the resources or the knowledge to cover properly. Nevertheless, in 1978 we began to investigate the markets in Iran and Saudi Arabia. I recall one trip to Tehran, while the Shah still reigned, when I spent most of 10 days waiting for meetings with various government and military officials. Because of the unproductive time that I spent waiting, it was a frustrating trip. It also proved to be an unsuccessful trip.

Some parts of the world required a very large marketing investment. We did not have the financial or human resources required to penetrate those markets. In 1982, after a visit to China, I put that country in the same class. Part of our problem was that we did not have representatives in those countries. We made several visits to Moscow, when it was still part of the Soviet Union; we had a knowledgeable representative there to help us find our way through the bureaucracies. The result was business with Aeroflot and today business in many countries of the former Eastern Bloc.

Though we only had four people carrying out our front-line marketing, we continued to develop a general aviation business by making effective use of our network of fixed base operators. Our small team went everywhere, including all the major trade shows, conferences, and industry association meetings. We exhibited at the Farnborough, Hanover, Paris, and the U.S.A. National Business Aviation Association air shows, as well as several smaller events. We learned how to write advertisements and carry out a trade magazine advertising campaign.

We did all of that work with a marketing budget of less than 5 percent of the product's gross revenues, though our investment in the product line's research and development remained high at more than 15 percent of gross revenues.

In a 1978 year-end product review, I reported on 20 development activities that we had under way, and those were only the ones that I considered important enough to review with our corporate management. Two of them involved new product ideas, and several were major new features for our existing Omega Systems. During the year we had solved a host of large and small technical problems; however, we still had several problems outstanding. That period was not easy for our engineers. Fortunately, everybody understood that if we were going to survive in this market, we had to work together and hard.

By end of 1978 we had more than 20 airline customers in 12 countries. Our airline customer base and the countries in which we were doing business doubled during the next two years. We had built an $8 million business, and it was still growing.

When we closed the financial books for 1979, our commercial avionics products business had grown to 20 percent of the division's business, which also had grown. By 1980 our commercial products represented 25 percent of the division's business.

More Initiatives, 1978–81

The first time that I had enough knowledge and experience to produce a comprehensive long-range strategic marketing plan was early in 1978. Six months later, shortly after I assumed the dual responsibilities of Omega product line manager and commercial aviation marketing manager, I wrote a short-range tactical commercial aviation marketing plan and a set of goals for our team.

Several such plans and surveys had preceded these plans, but the earlier ones had focused more on the military markets. It was not until we had experience, and until we had met with many potential customers, that we could produce a sensible worldwide assessment of the civilian business.

Our assessment included an analysis of the airlines and military transport air operations, the available navigation and cockpit equipment, and its capabilities. We studied the airframe manufacturers' marketing strengths and principal markets. We talked and listened to many military and civilian operators, not just their pilots, but also their engineers and accountants. We found out what they wanted and what they could afford.

Some manufacturers make the mistake of expecting their prospective customers to tell them precisely what product they want. Most users of technological products do not describe a product idea in detail, nor should we expect them to do so. Customers can describe what they need to make their operations more effective. Then it is the job of the manufacturer to translate those needs into a product.

Our market assessment included a thorough understanding of our competition. We listed the capabilities of our competitors' products and made our own assessments of the companies' strengths and weaknesses, their marketing, and, based on previous bids, their pricing strategies. Also, we tried to estimate what their future product and marketing strategies might be.

We then compared all that information with our product line and an honest assessment of our strengths and weaknesses. Once we had all the necessary information, we were ready to develop a product development and associated marketing strategy.

Through that process we identified several new products and new features for our existing products. Also, we were able to identify plausible markets for our company.

Until 1978 our Omega marketing had concentrated on those customers that needed an over-water navigation system. When the FAA allowed airplane operators to use Omega Navigation Systems as the primary means of navigation, our market expanded to include domestic operators. Because other countries followed the U.S. lead and adopted similar rules, we had prospective customers in most countries.

A common requirement among our first customers was their need for a standalone navigation system. Those of us who understood our market realized that it might remain stable, but more likely it would taper off. We either had to take the initiative to expand our market or accept the results of a declining business base.

The Boeing 747 jumbo jet used Inertial Navigation Systems for navigation, attitude control, and heading information, three essential pieces of data that only a gyro-based system could provide. The Omega System could only provide one of the three—navigation. Other systems on our customers' aircraft provided compass and aircraft attitude information.

Despite that shortcoming, Omega equipment had two advantages over the systems that used gyros: it was more accurate and cost less. The accuracy of an Omega Navigation System remained the same, regardless of the duration of the flight, whereas the position of a gyro changed over time, resulting in a gradual decrease in the accuracy of an Inertial Navigation System. The early airline gyro-based Inertial Navigation Systems degraded by about 2 nautical miles for each hour of flight, and so after two hours an Omega Navigation System was more accurate.

The other problem was cost. Airplanes had to have two working long-range navigation systems before they could enter the oceanic airspace. If an airplane leaving Chicago had only two Inertial Navigation Systems installed, and one of them failed before the airplane reached the coast, it would have to return. To safeguard against that occurring, most airlines installed three Inertial Navigation Systems in their aircraft.

When the regulatory agencies allowed airlines to use Omega equipment as a primary means of navigation, we realized that we could save the airlines money and improve the navigation of their airplanes. We proposed that the airlines install an Omega Navigation System instead of a third Inertial Navigation System. We also suggested that we connect the systems together so that the more accurate Omega position could be used to update the Inertial Navigation Systems.

We discussed that idea with Pan Am, and it was receptive. It was up to us to take the initiative. We were trying to respond to the dual need within the community to reduce costs and to improve navigation.

Our other initiative was primarily for business aviation. Our analysis of the market confirmed that business aviation, airlines, and military operators were becoming increasingly interested in flight management systems. Those systems relieved pilots from continuously controlling and assimilating information from the variety of navigation and communication systems on aircraft. Our earlier plans for an Omega sensor that could be connected to a flight management system now became more firm.

By working with the manufacturers of those management systems, we could design an Omega System that did not need its own control and display unit. Of course, such a product would not sell for as much as a full Omega Navigation System. However, by being responsive to our customers, we had the chance to expand our market rather than to watch our principal market diminish.

Consequently, we decided to take the initiative to expand our market on two fronts. We pursued the market for a combined Omega and Inertial Navigation System and an Omega sensor to work with the new flight management systems.

Some might call that market push. However, I have always thought that term connotes the supplier trying to influence the market. Successful companies respond to market needs; they do not try to influence their market. I prefer to call my version of this type of business development practice market initiative. In that process, the supplier first takes the time to understand what the customers want and then translates that knowledge into a responsive product. Next, the supplier takes the initiative to be the first to respond.

In his book *The Borderless World*, Kenichi Ohmae, the Japanese managing director of the consulting firm McKinsey & Company, describes his experience with a Japanese home appliance company. The appliance company was about to develop a coffee percolator to add to its product line. It started its development planning by looking at its competitors' product to see how the company could improve on the standard design—smaller, lighter, cheaper. Kenichi Ohmae convinced the company to talk to coffee drinkers to find out what they wanted and not to look at existing coffee maker designs. It found that what coffee drinkers care most about is good-tasting coffee. They do not care what the machine looks like or how it works. Therefore, the company's engineers worked to find out what influences the taste of coffee. The result was a machine that dechlorinated the tap water poured into it, heated the water to just the right temperature, and then regulated the water flow through the coffee that the machine had freshly ground.

To launch our integrated Omega and Inertial Navigation System concept, we first had to find a prospective customer that was willing to be our guide. Once again the commercial aviation pioneer, Pan Am, stepped in to help. It was in the market for a new airplane and wanted both types of navigation systems installed.

Delco and Canadian Marconi had already supplied equipment for Pan Am's other airplanes. Therefore, it was reasonable for us to expect that we would be the suppliers of the navigation equipment for its new aircraft. As a result, we held more detailed and frequent discussions with Delco than with Litton, the other supplier of Inertial Navigation Systems. We worked with Lockheed, Boeing, Douglas, and Airbus Industries, the four companies competing to supply Pan Am with new airplanes.

With the help of people at Pan Am, such as the venerable Pat Reynolds and Jacques Raia, a system concept started to emerge. We involved the FAA in many of our deliberations because we would be asking it to certify the new Omega and Inertial Navigation System combination. Once we were confident of the approach we should take, our engineers went to work to make it a reality.

Boeing remained committed to the Inertial Navigation Systems for its B-747s and to our surprise was reluctant to make any change. It agreed to provide Pan Am with Omega and Inertial Navigation Systems in its B-747, but it stopped short of integrating the two systems. Douglas and Lockheed were more flexible.

By the end of 1978 Pan Am had selected the Lockheed L-1011. In 1979 we flew a prototype version of a Delco Inertial Navigation System working with a Canadian Marconi Omega Navigation System. Then, in 1980, Pan Am put the first of its new L-1011 aircraft into service with the systems. It was a success from the beginning.

Varig decided to use the Litton Inertial Navigation System and the Canadian Marconi Omega Navigation System working together on its DC-10 airplanes. Varig's decision brought Marconi and Litton together for a joint development program. Making the two systems work together was an easy matter compared with making the Omega System work on Varig's DC-10 South Atlantic routes, when Varig again changed its flight schedule by one hour and we uncovered another Omega signal problem. Varig suspended our contract until we solved the problem, but fortunately we had a good working relationship with the airline. Varig's management understood that complex avionics sometimes require modifications to make them work in new situations. It agreed to leave our Omega System installed on the DC-10 so that our engineers could investigate the system's performance under actual flight conditions. Once again, Varig's professionalism and leadership saved unnecessary expense for the airline and our reputation.

Varig became increasingly anxious for results. Every week we sent an engineer to Varig with new software, but our rush to find a quick solution to the problem soon became counterproductive. Finally, I asked Assis Arantes if we could have three months to investigate the problem. I wanted our engineers to focus on the problem and not on quickly producing new versions of the software in the hope that we had cured the problem. Assis agreed to the moratorium. Sometimes less haste really does mean more speed and a greater chance of success.

Finally, we found that at certain times signals that had traveled 18,000 miles around the world were stronger than the same signals that had traveled only 6000 miles in the other direction. That variation occurred only near the equator, at certain times during the day, with signals from only one of the eight stations. The situation defied all of the most advanced theories about very low frequency signal propagation. The system's computer, thinking the signals had come over the short distance, applied the wrong signal correction factors. The solution was straightforward once we understood what was happening. When described in just one paragraph the problem appears to have been straightforward. However, as with many of the Omega signal problems that we uncovered and solved, their simple description, once we knew the problem, disguises their complexity.

Several other airlines and many general aviation customers saw the benefits of a combined Omega and Inertial Navigation System installation. In 1980 we took that concept one stage further. The Flight Systems Division of the large Sperry Corporation in Phoenix, Arizona, had produced an inertial attitude and heading reference system that used a new, lower-cost inertial technology. Sperry's marketing staff thought there would be a large market for such a system if it included an Omega navigator within its main electronics unit. Sperry asked us if we would be interested in a joint development project. We thought, as Sperry did, that there should be a large market for such a system, and so we started the design work while we negotiated a joint agreement. Interim agreements protected both parties.

This was one of our first joint development projects with a company that was not part of the General Electric Company conglomerate. The arrangement showed that the relative sizes of the companies in a joint project need not be an obstacle to cooperation.

Sperry's size did not intimidate us, though we were flattered when it asked us if we would join it in a joint development project. Our agreement was straightforward. It established Sperry as the lead for the project. We agreed that each party would fund its own design activities and that we would help Sperry with the system assembly, flight test, certification, and marketing. Each party retained the manufacturing rights for its own parts of the product. We also agreed, at the beginning, on transfer prices and selling prices to make sure that one party could not later hold the other for ransom. The agreement included the protection and restrictive use of the technology that we would exchange during the project.

Pricing, proprietary data protection, and similar agreements are necessary in any joint development, but they are not the chief ingredient, as some think. Joint ventures succeed only when the parties trust each other, and respect each other's contributions and when each party is making an important contribution. Joint ventures should be equally beneficial. There is no room in a joint venture for a freeloading partner.

In our arrangement with Sperry, we had the all-important Omega technology and the reputation as a leader in the Omega market. Sperry had considerable expertise with inertial systems and was a major supplier of aircraft electronics systems. It also had a large worldwide marketing and product support organization.

Some companies can be very protective of technology in the belief that this secretiveness alone will ensure their continued competitiveness. Companies keep, or improve, their positions in a market by responding to customers' real requirements and continuously improving their own processes and technologies. Protecting current or outdated technologies will not guarantee continued success.

A prototype version of the joint system was ready for its first flight test in just six months. Preproduction equipment was ready within 12 months, and we were ready for production within 24 months of signing the original agreement, which was a remarkable achievement. Sperry management committed considerable resources to the effort and convinced our senior management to do likewise. We were always impressed, and a little intimidated, when Sperry senior managers came to a meeting in Montreal in their own corporate jet; we did not have one.

To supplement our already overworked Omega engineering team, we hired several experienced engineers. From other companies, we enticed people such as Jim Bruce, Tony Murfin, and Sylvia Sayegh. They brought significant hardware and software experience to our team. One of the early junior members of the Omega team, Jean-Pierre Lepage, was now a senior software engineer and marked for further promotions. Those people have remained crucial members of the Canadian Marconi commercial avionics team.

The engineers at both Sperry and Canadian Marconi had conceived a system that solved several problems for our prospective customers. Principally, it provided an inertial-type navigation, attitude, and heading reference for half the cost of conventional gyro-based navigation systems.

Our joint development of an Inertial-Omega Navigation System, as we called the new system, was a technological and joint development project success but a market disaster. The product was our idea, not one suggested by the airplane operators. It was a classic case of market push: a manufacturer believing it could convince the customers that they needed the product. Our separate but connected Omega and Inertial Navigation Systems concept, which we had successfully designed and sold, was the direct result of responding to customers' requests. In that case, we had exercised market initiative.

Two crucial factors led to the demise of the Inertial-Omega Navigation System that we developed with Sperry's Flight System Division. Inertial Navigation Systems were mature and certified. Also, Inertial Navigation Systems were in widespread use. The conservative airline industry saw no reason to take a risk with a novel concept. The airlines and their airframe manufacturers had invested years of effort in bringing the conventional Inertial Navigation System to its present state of maturity and widespread use. Boeing, Lockheed, and Douglas were using those systems on their wide-bodied airplanes. Because our Inertial-Omega Navigation System was much less expensive, we targeted the Boeing B-727 program as our launch customer. Boeing, though, was already considering fitting the Inertial Navigation Systems.

It was easy for Boeing to fit the standard Inertial Navigation System. The installation, technology, and certification requirements were well known. If it fitted the Sperry-Marconi Inertial-Omega Navigation System, it would be dealing with a new technology and an entirely new certification program, the cost of which would cancel most of the benefits of our system.

By that time Boeing had launched the development of two new aircraft, the B-757 and B-767. It expected to use an even newer technology for the inertial systems on those airplanes. Honeywell and other companies were developing inertial systems that did not have any moving parts, like gyros, but employed laser technology. We hoped that the early problems that the ring laser gyro developers were having would give us an advantage. It was not to be. The persistence and competence of companies like Boeing and Honeywell paid off. It was not long before the aviation community had a working ring laser gyro Inertial Navigation System. Our inertial-omega market did not develop; we had not listened closely enough. Fortunately, our core Omega products and the new feature that allowed us to connect existing Inertial Navigation Systems with our equipment were selling.

While we were working with Sperry and taking care of our core products, we started the design of an Omega sensor to work with navigation management systems. We intended to assign an entirely new system number to the Omega sensor, though it was a derivative of our successful CMA-734 Omega Navigation System. At first, we rationalized there was a marketing advantage to announcing a new product. Then we realized that we could not capitalize on the good name of the CMA-734. Instead of announcing the CMA-759, itself a sterile name, we announced with much fanfare the CMA-734 sensor. Owners of the standard CMA-734 could easily have their systems modified, if they wanted them to work with a navigation or flight management system that managed more aircraft systems than the early navigation management systems. Also, we could take advantage of much of the certification work that we had done for the FAA, the British Civil Aviation Authority and many other regulatory agencies throughout the world. We had applied another marketing lesson we had learned during our search for excellence. Sometimes it is prudent to maintain a product with variants rather than to make it obsolete by introducing a new similar product.

To further expand the market for the CMA-734 and to reduce its cost, we began to design a simpler control and display unit. In 1980 we talked to several airlines to find out how we could improve the usefulness of our airline Omega Navigation System. To that point, the CMA-771 and navigation systems like it could store only nine waypoints. After the aircraft had flown over those waypoints, the pilot had to insert the next set of waypoints on the route. To avoid all that work, some airlines and business jet operators were starting to install the new navigation management systems that could store many waypoints. Many customers asked us if we could expand the capability of our Omega equipment to provide a similar feature.

The technology was straightforward, although we did not know how we could cope with or fund more design work. Already we had the project with Sperry under way, we were designing an Omega sensor and a small control and display unit, and we had other projects, all designed to expand our core market.

We wanted to be sure that we understood what our customers wanted, and so we spent a year discussing several options with our major airline customers and the airframe manufacturers. No other Omega manufacturer was responding to this multiple waypoint requirement. We had a little time to be sure of our approach and to complete several of our other equally important projects.

In 1981 we started to work on a major new upgrade to our CMA-771 airline Omega Navigation System, which became known as the CMA-771 Alpha-Omega. By then we had learned to name our products because that helped to differentiate them in the market. Using the name Alpha allowed us to title advertisements with phrases such as: "Alpha-Omega, The First Word in Long-Range Navigation from Canadian Marconi." The name also indicated the most important feature of the system. As well as identifying the waypoint simply by a number, our system displayed the waypoint by its internationally recognized code. For example, when the system was on the final leg to runway 24 left at Dorval International Airport in Montreal, it would display YUL24L.

The system that we designed had the capability to store 500, and later 4000, waypoints, all sorted into an operator's regular routes. We replaced the rotary switch, which the pilot had used to select navigation data on the first control and display units, with nine push buttons. Those buttons in turn, gave the pilot access to much more information than had been possible before.

In the late 1970s, when we were still trying to expand our market, we used our Omega technology in many other ways. The Canadian Armed Forces maintain a presence in the Arctic, in part to underscore Canadian sovereignty of our northern region. Canada's frozen north presents a unique challenge for navigators. Today, satellite-based navigation systems have solved most of the problems, but those systems have only become available recently. Before satellite navigation, navigators in the Arctic and other remote regions, like the Sahara Desert, had to rely on compasses and celestial navigation.

We worked with the Canadian forces to produce a variant of our small CMA-734 Omega Navigation System that would work in the vehicles they were using for travel in the Arctic. They conducted two sets of tests during a two year period. Although the tests were successful and the Canadian forces bought a few systems, we did not find a large market for that land navigation system. We failed, I believe, because we were venturing outside the aviation market that we knew so well. We were an avionics supplier, accustomed to dealing with aircraft operators, not the operators of large, all-terrain, long-range vehicles. Many technologies are adaptable to a variety of products, but diversification into entirely new markets, as we were finding, can be a difficult undertaking.

In the early 1970s our Avionics Division had designed a satellite-based positioning system for ocean and land survey use. Although the product worked well, we were never able to secure enough business to make the product a commercial success. We did not understand the support requirements of that industry. Also, we did not find a cost-effective way to customize, as the survey companies expected, each system we sold. Our avionics products had modest batch production runs, and so we did not have to produce them one at a time, as we found we were doing for the satellite survey equipment.

Another initiative to expand our Omega market failed because it relied on a military initiative that did not receive funding for the production phase. After the Vietnam War, during which many aircraft on survey missions had been shot down, several countries began to develop remotely piloted vehicles. They were designing those vehicles to carry out missions that could be done with cameras and electronic surveillance devices and without crew.

Before it became possible to navigate using signals from specially designed satellites, the navigation of those remotely piloted vehicles was a problem. The Omega System presented one possible solution. Working with a company that planned to offer a remotely piloted vehicle, we designed a special, very small Omega navigator for the vehicle. Before we could reap the benefits of our work, the program was canceled, or so we were told.

Sometimes circumstances beyond a company's control will stop an initiative to expand a core market. That is the price of business. Successful companies learn from those experiences but continue in their quest to expand their market base. We became very imaginative in our quest for new markets.

Many short-range aircraft like the B-737 do not need a long-range navigation system except to get them from the Boeing plant in Seattle to a country thousands of miles away. We designed a carry-on Omega Navigation System using our small CMA-734 in a case. The kit included cables so that a technician could quickly connect it to the essential airplane systems and an antenna. Boeing tested our first carry-on Omega System in 1979. Later, it and other airframe manufacturers bought several of those kits for aircraft deliveries. It was a modest initiative, but it illustrates how we responded to real needs throughout our market.

I was disappointed when I was unable to convince Canadian Marconi to support a customer financing initiative. I tried to introduce a trial purchase program. It was an attempt to attract more orders from the users of corporate aircraft. My idea was simple. The program required a customer to sign a contract whose payment terms were net 90 days, not 30 as we usually stipulated. I worked out a price that considered the extended payment period. The proposal also limited the number of systems that we would offer, at any one time, to limit the company's risk investment. I reasoned that very few customers would go to all the trouble to install a system and then return it after 90 days. In many ways, my suggestion was a marketing gimmick, and I suppose that was how others in the company viewed it.

In 1983 I tried to introduce a leasing program. That also did not gain acceptance within the company. We were a manufacturing, not a leasing, organization.

Despite our few failures, most of our initiatives to expand our market were successful. The most successful ones focused on our core business and products: the airline, business aviation, and military markets and products that responded to the evolving requirements of our expanding customer base.

By being ever conscious of our customers' needs, we managed to build a solid commercial avionics business. That base allowed us to launch successfully several entirely new commercial avionics. Then we only had to sell the product. The company's capabilities and commitment were no longer a question. We had a track record.

16

"Profit" Is Not a Dirty Word

You cannot keep out of trouble by spending more than you earn.
—Abraham Lincoln

Another Lesson from Pan Am

During the final stages of our negotiations with Pan Am, we learned another important lesson—profit is not a dirty word. Our prospective customer was the unexpected source of that revelation.

When Pan Am realized that we had not included enough funds in our bid to set up a local repair center for our equipment, its negotiators gave us a lecture. In part, they told us: "To be successful, a business arrangement must be profitable for both parties." Pan Am had had the experience of a supplier, desperate for the business, selling it equipment at or near cost. Later, when Pan Am needed help, the company was not anxious to spend more money on an already unprofitable contract. As a result, both parties suffered.

In an unprecedented move, Pan Am asked us to go away and revise our bid. There we were, negotiating a contract that would dictate the success or failure of our product, and the customer was inviting us to increase our bid.

After much debate we decided to contract with Butler Aviation in New York to provide Pan Am with around-the-clock support. We amortized the cost of setting up that service over our expected U.S. market. We increased our bid, in proportion to our amortization strategy, to give Pan Am what it wanted, and a few weeks later it awarded us the contract.

Our strategy was to take a longer-term approach to the civilian market. Our goal was to make a modest near-term profit with the expectation that our longer term consolidated profit would be at or near the company's overall profit goal. We were well rewarded for not being greedy for near-term profits.

Most management theoreticians condemn the concept of management by numbers. A common mistake is to define management by numbers as a management technique that bases business decisions solely on achieving a given return on investment.

I find it difficult to believe that companies base decisions solely on expected profit margins. Those that do are employing a technique that I prefer to call management by profit. Companies, like families, must make a profit to avoid crises and expand their activities. However, we must all make judicious investments to achieve long-term sustainable benefits. The most obvious example for families is the sacrifices they have to make to buy their first homes. Successful companies understand the need to make sacrifices and to lower their short-term return-on-investment expectations when they are entering a new market. Like a family's first home, a new market requires a long-term investment approach. However, also like the family with a mortgage, the investment should not be so large that the company cannot recover the investment or the family cannot pay the mortgage.

I believe in the notion of lowering short-term gains to achieve long-term sustained success. I am not a proponent of unprofitable contracts. Management by profit usually only produces short-term gains. Companies that have operated in that fashion have conceded entire markets and, in some cases, their entire business.

I call my version of management by numbers management with numbers. That phrase is more accurate because it suggests that numbers, financial, and inventory statements are only ingredients in the decision process. A management decision must also take into account market conditions, customer requirements, the company's long-term goals, and several other factors.

Management with numbers requires current data. Semi-annual, even quarterly, reports are not sufficiently current for most decisions in a manufacturing environment.

Any day it was an easy task for us to know the financial health of the company or a particular product line. We believed that managers should have ready access to all the tools that they needed to do their jobs, which included current financial and inventory data.

Each month managers received a report on the financial and labor hours status of their cost centers. Product and manufacturing managers received a record of each factory order and the inventory for which they were responsible. Other statements provided information about that month's shipments and order bookings.

We frequently complained when each month's computer printouts buried priority paperwork on our desks. However, those data helped us to run our business effectively.

The factory order statements allowed us to keep track of the costs for each step in the manufacturing process. With that knowledge, we could search for ways to make the most costly steps more efficient. Current knowledge of our engineering, marketing, and support costs helped us to plan our work. With a measurable plan, we did not have to frequently waste time on corrective activities. We "planned our work and worked to our plan"—most of the time.

Inventory statements were critical to our efforts to stock only those parts that we needed for the number of systems we wanted to build in the near future. The phrase today is just-in-time inventories.

As a supplier to the military, we had not worried too much about inventory management because most of the time we only built equipment after we received an order. Then, delivering the equipment on schedule and within budget was our primary concern. However, soon after we entered the civilian market, we realized that we had to build equipment in the expectation of orders. Then, if we did not secure the orders, we paid the penalty of excess inventory.

When we received orders, our customers often required deliveries within weeks or months. That was an entirely new situation for us, but we knew that if we did not offer prompt deliveries, our customers would go elsewhere for their navigation equipment. Because we were ever conscious and responsible for the financial state of our product line, we quickly learned the benefits of just-in-time inventories.

We used our knowledge of the financial state of our product line to help us with our decisions—not to control them. We were entering a complex and, for us, new market. We were designing and building an advanced aircraft navigation system for a variety of customers. Many of our customers had their own set of requirements. Some wanted off-the-shelf deliveries of the standard version of the product, whereas others were willing to wait until we designed and added a special feature. Still others conducted protracted flight tests of competing equipment, followed by the negotiation of a complex contract. It was essential that we maintained control of our engineering and production activities, which included their costs.

Costs can be a one-time expense, some of which I have already discussed; other costs are incremental expenses over many years.

When we ventured into the civilian avionics market, we found that our customers expected a different kind of contract than the military. In a military contract each obligation is listed and priced separately. For example, a typical military contract will include payment for any special design work, the equipment, spares, warranties, installation support, training, and manuals. Thus, the military organization expects to pay for and receive exactly what it lists in the contract.

By comparison, our civilian customers expected the equipment price to be all inclusive. *The World Airline Suppliers Guide* outlines all the services the airlines expect from their suppliers, regardless of whether they are buying a multimillion-dollar airplane or a few-thousand-dollar piece of avionics equipment.

To complicate matters, many airlines and airframe manufacturers produce their own versions of the guide.

The short version of that complex set of contractual obligations is that an airline expects various types of support for as long as it owns the equipment—at no additional cost. The only exception is equipment repairs after the warranty period.

That was the environment in which we found ourselves. We provided the same type of service to our first large customer, Pan Am, and our smallest, most distant customer, Sea Bee Air, which operated a Grumman floatplane in New Zealand.

If we had not managed with numbers, we could have easily bankrupted the product line. However, the numbers were only one ingredient among many that we had to consider when making tactical and strategic decisions.

Although we tried to use a standard set of contractual conditions to improve the efficiency of our internal review, we understood that we had to remain flexible. With the customers we changed and customized our commitments to cater to their special needs.

Complex technologies and contractual commitments were not new to the company. We were familiar with the varied product support requirements of the industry. However, some tasks, such as providing a 24-hour emergency service every day of the year, were new but manageable.

What was new to us was putting the entire package together as one all encompassing commitment that was paid for in the price of the equipment. The normal military warranty was one year. We broke with tradition and offered a three-year warranty with our Omega equipment. That had two effects: it indicated to our prospective customers our confidence in the product, and it sent a message to our staff that we were serious about product quality.

A company can gain a positive reputation for product support. However, a product should not need a lot of after-sales service if the manufacturer builds it right the first time. Customers would prefer equipment to work as advertised and to continue to work; they would rather not have to call on the supplier for service. The ideal after-sales service is one that caters to a customer's changing requirements, not one that just fixes broken equipment.

As we entered the civilian business, we found that we had to project our cost of sales over many years. It was no longer enough to estimate only the short-term costs associated with a sale.

Development investments are part of the cost of sales, but the most difficult item that we had to account for was the cost associated with providing the comprehensive product support program that the airlines required. Unlike the warranty costs that covered a three-year period, the product support program could continue for 20 years.

We became technofinanciers, while we tried to estimate the costs of that service, because we had to amortize those costs in the price of the equipment. If we were too conservative, our product's price would not be competitive. If our estimates were too low, we would quickly get into financial difficulties.

By 1978 we had learned how to build equipment on risk, that is, investing in production without a firm order—an abnormal procedure for a military supplier. We developed procedures for just-in-time inventories to lessen our risk and investment. With the help of customers such as Pan Am, Boeing, and Varig, we began to understand the complex support requirements of the civilian market. Once we understood, we moved fast to put in place the financial and support infrastructure that was necessary to meet the needs of our new markets. To provide our customers with prompt service, we had set up three repair centers. We developed long-term pricing policies, and we were routinely agreeing to contractual commitments that only two years before we had thought to be too risky.

Our product and its market were complex. By North American standards, we were a small company, and we were competing in the big league. We listened, we learned, we acted. Then we succeeded.

None of our successes would have been possible without initiative, courage, and an understanding of our costs. Our practice was, in part, to manage with numbers the essential but not the only ingredient to success and profitability. Profit is not a dirty word; it is just another word for a positive balance of payments, without which a business will falter.

The Customer's Rights

*Any activity becomes creative when the doer cares
about doing it right, or better.*
—John Updike

Profit means having customers, and a company may have excellent products but still fail. Customers have several rights that a supplier must respect. Many companies believe that if they produce a quality product, they have fulfilled their obligations. Then they expect the marketing department to make the product a business success. Those companies are wrong. Customers have the right to receive a quality product, but that is the easiest of the customer's rights to fulfill. Matching quality with a competitive price is more difficult. Suppliers of goods and services, if they expect to succeed, must pay attention to all the needs of their customers. Those needs include product performance, after-sales support, and the evolution of their products to satisfy continually their customers requirements.

Customers have two fundamental rights: the right to be heard and the right to receive what they paid for. That means suppliers have to listen and to provide what they promised. Fundamental rights require money.

Listen... Listening can be an expensive process. In the early 1970s, we offered our first-generation Omega Navigation System, the CMA-719, and our only real customer was the Canadian Armed Forces. Without their early support of this new type of navigation and our product, we may have failed even before we started. A long time passed before the airlines embraced the Omega System, and when they did, they considered our first product to be obsolete.

We talked to the operators of small aircraft and helicopters. What we heard was that the available Omega Navigation Systems were too big for their aircraft. We took the initiative in 1974 and responded with the CMA-734.

While we were designing the CMA-734, we listened to the airlines. By 1975 we were completing the design of the CMA-740, which conformed to the first industry specification for an airline Omega Navigation System.

As soon as we started our negotiations with Pan Am their pilots asked for changes to our design. Those changes were easy for us to make, but other changes and new product features that we offered in response to customers were not so easy. Soon after the first airborne Omega equipment went into service, customers found ways to improve the product.

In the mid-1970s, when we started to sell our second-generation Omega products, the Omega signal coverage was not always adequate. We responded by adding a receiver so that our equipment could use the signals from the U.S. Navy's very low frequency communication stations as a back up form of navigation.

One of the reasons for our early success with Boeing was that we listened. Working with Boeing engineers, we designed an amplifier and signal splitter, so that the Omega equipment could share an existing antenna. That small, inexpensive device reduced the installation cost and helped us to secure Boeing as a customer. The device also became useful for several other customers.

When we introduced our next airline Omega System, the CMA-771, in 1979, many more variants resulted from listening to customers. We designed some systems to work with the Litton Inertial Navigation System and others to work with the Delco Inertial Navigation System. In response to users of search and recovery aircraft, we provided an optional software program so that the system could guide the aircraft in prearranged search patterns. Some of those operators wanted an output of position to a surveillance camera in the aircraft. For Boeing, we designed interface circuitry so that the system could provide information to a Performance Data System that was unique to its B-727s. Still other customers wanted the Omega equipment to send signals to the weather radar equipment or other displays in the cockpit.

Flexibility kept our entrepreneurial spirit alive. As a result, our entire product team felt, and rightly so, that they were making a contribution to our continued success.

Our willingness to listen and to respond to our customers gave us an advantage over some of our competitors.

When the airlines told us that they wanted their navigation systems to store more than nine waypoints, we responded with the CMA-771 Alpha-Omega. We started its development in 1981. Through discussions with several airlines, we came up with a design that the pilots wanted. To be sure of our approach we made prototypes of several possible versions of the pilot's control and display unit. Then we asked the airlines for their opinion. When we introduced the Alpha-Omega to the market within two years, it was an instant success because we had listened.

Our existing customers asked us to come up with a CMA-771 Alpha-Omega design that would allow them to upgrade their existing equipment, and so we did. In fact, a customer with our first airline Omega Navigation System, the CMA-740, could upgrade it to a system that could store 4000 waypoints and that had the very low frequency receiver for backup navigation purposes, as well as features that allowed it to connect to other aircraft equipment.

Designs that allowed customers to upgrade our older products contributed to our success. As we included the latest electronic technologies—smaller components—space became available in the main electronics unit for more circuits. For more than 15 years, during which time we introduced many new features, we strove to ensure that we did not make the earlier systems obsolete. With that approach, we kept our customers and attracted new ones.

When we did not listen, we usually failed. The integrated Inertial-Omega Navigation System that we developed with Sperry's Flight System Division in the early 1980s failed because we did not listen. We tried to push the market, even when the market was satisfied with the gimballed gyro Inertial Navigation Systems. We chose to ignore the signs that the airlines were looking toward using the laser gyro system that Honeywell and others were developing. We thought that the laser gyro technology was too risky and would probably fail. We had another solution, but no one was listening—neither were we.

Suppliers of products with complex technologies have a tendency to think that they know more about technology than their customers. Users of technological products are becoming increasingly better informed, and so today such an assumption is not necessarily true. Regardless, customers know better what they want. It is the job of the manufacturer to convert requirements into a product that customers want. That is the art of listening.

Manufacturers have a responsibility to educate their customers about their products and then to be responsive to their calls for help. We learned to take every call for help seriously and to be responsive to our customers' many needs. A quality product is only one ingredient for success.

Also, manufacturers should participate in all aspects of their industry. We spent many hours and traveled to many countries, helping to write the various Omega specifications used by the industry. In 1974 I helped to form the International Omega Association. In 1981 we hosted the sixth annual meeting of the association in Montreal. At that meeting I presented a paper on the return on investment of Omega equipment. Afterward, we made many such analyses for airlines. Those analyses helped them decide if Omega equipment or another form of navigation was the right choice for their particular situation and route structure. That was just one of the many services that we provided to help the industry. Of course, we had a marketing motive for carrying out those studies. However, like most services of that type, we could not directly relate the investment to its contribution to our revenues. Those services were an expense that, as those who have a commitment to an industry know, was necessary for success. Successful companies also put in place, as we did, a comprehensive product warranty and support program.

Provide what you promise . . . At its 1980 convention, the Aircraft Electronics Association elected Canadian Marconi company as its manufacturer of the year. The citation on the plaque read: "In recognition of their outstanding contribution to general aviation electronics for the year 1980." The Aircraft Electronics Association had compared us with all the U.S. general aviation avionics manufacturers, big and small, as well as companies in Europe. Eight years earlier a voice from the back of the room had asked me: "Macaroni, wasn't he the Italian guy who invented the radio?" We had come a long way in eight years in making the transition into the civilian market.

We tried to promise only what we could deliver. It took us several years to become confident in this new, for us, market. Bill King of Pan Am was right, but also true to his word, when early in 1977 he said to Ervin Spinner, our marketing manager: "You know, Ervin, Canadian Marconi is really quite ignorant about the airline industry, but we are going to help you learn."

Through Boeing, Pan Am, and our other early customers, we started to appreciate the complex support requirements of the airlines. Setting up repair centers was a straightforward matter. That just took money, which we did not have much of, and an ability to select the right strategic locations. By 1981 we had 11 service centers in as many countries. One of our most significant achievements in that area was our successful initiative to negotiate repair center agreements with some of our airline customers. Four of our airline customers were offering a repair service to our other airline customers.

Our three-year warranty was one of the best in the business. Other than obvious abuse, it was a warranty with few restrictions.

The reliability guarantee was our first real support challenge. In the ideal situation, a product does not fail. However, even the best products fail, especially technological products operating in an environment like an airplane. Our airline customers expected us to declare the failure rate of the product and to back up our claim with a guarantee. Airplane operators needed serviceable equipment, and so their solution was to make their suppliers loan them equipment when a product failed to meet the guarantee.

The reliability guarantee included a formula that the airlines used to calculate how many spare units we would have to lend them at no charge. The formula used information on the number of times the equipment had failed during an agreed period and how many systems the customer had bought.

We learned quickly that we had to accept those risks if we expected to survive in that market. The reliability guarantee also provided us with an incentive to produce, and to continue to produce, a quality product.

Electronic products have a greater tendency to fail earlier, rather than later, in their service life. To reduce those early failures, we employed three processes. First, through training and the use of quality production tools, we helped our production staff to maintain a high standard of workmanship. Although we had inspectors, we did not rely exclusively on postassembly inspection as a way to product quality. We subjected all electronic parts to electrical and thermal stress tests before the assembly line staff used them in production equipment. That way we identified and discarded any weak components.

As an extra precaution we subjected every unit that came off the production line to a 120-hour test cycle. The test included heating, cooling, and vibrating the equipment while it operated. By the time we had an Omega Navigation System ready to ship, we were confident it would not fail for a long time.

Another aspect of our support program that initially gave us cause for concern was the emergency service. In an emergency, our airline customers expected us to respond to a call for help within four hours. Normally, they would only declare an emergency if they had a problem with our equipment on an airplane that was ready for departure. We did not have to fix the problem within four hours, but we did have to initiate a solution within that time. That might mean sending an engineer or a spare system on the next available flight. Our customers expected us to provide that service 24 hours a day every day throughout the year. Calls for help could come from anywhere in the world.

To cope with that situation, we set up an emergency telephone service and a duty roster. Most of the people connected with the Omega product line had a duty officer schedule. Most of us also gave our home telephone numbers to our customers. Such was the spirit of our team.

We dreaded receiving the call declaring an AOG, airline code for "we have an aircraft on the ground." Fortunately, such calls came infrequently. We had less than 5 such calls in 10 years. Airlines in particular will try to solve a problem themselves. They position spare systems throughout their route structure or borrow a similar system from another airline.

Once, I received an AOG call on Christmas Day. I called Hans Koller, manager of our Repair and Overhaul Department, who in turn called his senior Omega System technician, Gilles Bercier. We all met at the Canadian Marconi plant, and while Hans and I provided moral support, Gilles checked out a replacement system.

When the equipment was ready, we realized that we did not have the necessary shipping papers. We telephoned Gaston Roussel, manager of our Shipping Department, and he came to help. Once we had the system packed and the shipping papers in order, I drove to Montreal's international airport, some 40 miles away in a snowstorm. Our customs broker, whom we had also called out, met me at the airport. With little time to spare, our shipment was on the first leg of its journey to Fiji, where an anxious Air Pacific was waiting for the replacement equipment—or so we thought.

It had been a scramble, and those of us involved in the activity returned to our homes for what was now a cold turkey dinner. The following day we received a message from Air Pacific. It thanked us for the replacement system but also let us know it had solved its immediate problem by borrowing a Canadian Marconi Omega System from Air New Zealand. After all our work, and missing much of our Christmas Day, the emergency had not been as acute as we thought. Despite that nonemergency, we had shown, once again, that we supported our customers, regardless of their size or location. The experience taught us to be better prepared in the future.

Some airlines insisted on a written performance guarantee. A lay person would be right in believing such a request was reasonable. However, an Omega Navigation System's performance can be affected by events unforeseen by the designer. The radio waves that the receiver relies upon have traveled several thousands of miles, during which time various types of interference can change their pattern. In our world, phrases like solar flares, polar cap anomalies, ionospheric disturbances, attenuation, magnetic fields, cycle slips, and diurnal errors were part of our daily vocabulary. Each phrase represented a special event that we had to understand. All events had the potential to affect the performance of our system.

Frequently, a contract was a 70-page document that included the basic commercial terms and the product support program. We found out that the way to success was to back up our claims with real commitments. Customers expect their suppliers to provide what they paid for, and that means a lifelong commitment to the product and customer. At times, a supplier may have to take a risk and then make an investment when something unexpectedly goes wrong.

That was the case with my promise of 4-nautical-mile accuracy to Assis Arantes of Varig. Later, we confirmed that commitment in writing. After Varig had been using our systems in its B-707s and when it was about to install more systems in the DC-10s, our world fell apart. Unexpectedly, the system occasionally had errors as high as 100 miles. We spent most of a year investigating and finally solving that problem, but we had to make a large investment. If we had not lived up to our commitment by solving the problem, a bad reputation would have preceded us wherever we went. In that case we turned adversity into prosperity. We had showed that we always tried, regardless of cost, to honor our commitments.

All that and more can be accomplished profitably. An accurate and current understanding of the financial health of a business is an essential ingredient to success. With judicious management, by listening to customers and by giving them what they want and what they have paid for, companies can grow their business and show a profit.

17

Maintaining Momentum

*Listen, communicate, respond to needs and suggestions
of internal and external customers.*
—A Canadian Marconi Company
Continuous Improvement Principle

Between 1981 and 1983 we had more ideas for new Omega products than resources—both human and financial. However, we knew that if we were going to maintain our market momentum, we had to continue to be responsive to our market. During those years electronic component manufacturers also made significant improvements to their products, some of which I will discuss in this chapter.

The fiscal year-end report that I wrote early in April 1981 contained good news and justifiable optimism for the future.

Introduction
Revenues exceeded those predicted, three major new development projects were launched, an off-the-shelf delivery situation was achieved, product support was expanded, and new markets were developed.

Engineering
The year started with the launch of the CMA-860 receiver [the product for remotely piloted vehicles], by the midpoint the CMA-802 AHRS/ONS program was under way [our Inertial-Omega Navigation System joint project with Sperry] and toward the end we began work on an alphanumeric display unit [for the CMA-771]. In addition to these new development projects, several new configurations of the CMA-771 were completed . . . plus a host of software bugs [problems] were sorted out. During the year, we put considerable effort into resolving a Varig South Atlantic performance problem. This problem is close to resolution but only after many hours spent by product support

and [the Omega] engineering [team] in DC-10 flight decks collecting data on 10-hour flights. . . .

The launching of the CMA-802 [joint project with Sperry] caused an influx of many new engineers. All of these engineers have efficiently integrated themselves into the Omega team and have made a valuable contribution to the program.

Marketing

We exceeded the budgeted sales forecast [by $1 million] . . . despite the fact that there was not a single contract for more than 15 systems [we delivered 250 systems during the year].

Production

Omega production finished the year in style; during March 1981 just over $1 million worth of Omega products and services were shipped—well done!

Support

Our support image continued to improve during the year. We now have 11 service centers, inclusive of 4 airlines with third-party repair agreements, located around the world. The Field Support, Repair and Overhaul, and Training Departments were extremely active in all corners of the world . . . .

The Future

Our sales forecast for FY 81/82 is $7.6 million representing some 280 systems. . . . The Canadian forces alone will be buying 48 CMA-734 systems. [Other] contracts of note that we are working on include Air France, Pacific Western Airlines, Pan-Am/National [Pan-Am had recently bought National Airways], Canadian Pacific Airlines, and the U.S. Coast Guard. Our position with various general aviation aircraft manufacturers has improved, and we confidently expect the general aviation business to expand.

Conclusion

I would like to take this opportunity to thank everyone connected to Omega for their contribution during [FY] 80/81. Omega engineering, production, marketing, support, and all of the supporting departments worked conscientiously and hard at all times to make [FY] 80/81 a successful year.

I have liberally quoted from this year-end report because it highlights many aspects of our business situation.

At the time I wrote the report, we had 40 airlines as customers, among them Air Afrique, Air Guinea, Air Pacific, Air New Zealand, Avensa in Venezuela, China Airlines based in Taiwan, Cruzeiro and Varig in Brazil, Japan Airlines, Lan Chile, Lineas Aereas Paraguayas, Pan Am, American Transair, Greenlandair, Hapag-Lloyd in Germany, and Royal Brunei Air. Our 40 airline customers were situated in 25 countries on 5 continents. Seventeen military and government agencies had bought our Omega Systems, as had more than 100 operators within the general aviation community. Those customers, between them, operated 98 different types of aircraft, some of them helicopters, and each aircraft had presented a special installation challenge.

We had every reason to be optimistic about our future business because now we had products that our customers wanted and new Omega products in development that we were confident we could sell.

Usually, our predictions were accurate to within 10 percent, and so it was a pleasant surprise when we calculated our revenues that year to be 21 percent more than our forecast. Now, we were close to achieving our $10 million dollar per year goal.

That we achieved our goal within two more years was more remarkable because two of our new product ideas did not succeed, and the general aviation market went into a recession. The start of a cataclysmic downturn in general aviation airplane sales started in 1980. Some 5500 business airplanes were shipped to customers during the peak production year of 1979. Three years later there were customers for only 1000 new business airplanes.

The alphanumeric display for our CMA-771 airline Omega Navigation System was a success. Our engineers had to make several changes to the system's main electronics unit, including upgrading its processor and expanding the memory capacity. Once again, they designed those changes so that customers with our older systems could upgrade them to the new version.

As we started that work, we began to look further ahead. In a December 1980 Omega product line review with our management, I presented several ideas for new Omega products. I suggested that we design a control and display unit with a cathode-ray tube display and expand the capabilities of the CMA-771 Alpha-Omega so that it could function like a flight management system.

At the other end of the spectrum, we had ideas for smaller, but more powerful, Omega sensors to provide navigation information to flight management systems of other manufacturers.

To attract a wider market, we proposed to develop a simple, low-cost control and display unit that used a new liquid crystal flat screen display technology. With that unit, we could sell the Omega sensor also as a small, low-cost navigation system.

We looked at ways to improve the CMA-771 Alpha-Omega. Usually, that meant more memory, so that the system could store more waypoints. We also investigated a more powerful and faster processor, so that it could handle more calculations. We had ideas for more circuit cards designed specifically to connect the system to other aircraft systems and displays. Those features combined would give us our own flight management system, while still maintaining the market that we knew well—the Omega Navigation System market.

For eight years we had continuously developed new Omega products and significant new features to remain competitive. We produced more new designs than our competitors, and I am convinced that this continued investment in the product helped us to succeed. Another important contributor to our success was our decision to ensure that we could upgrade our older equipment with any of the new features we designed. There came a time, though, when we had to break away from the interrelationship between our small CMA-734 Omega Navigation System and our three airline products, the CMA-740, the CMA-771, and its successor, the CMA-771 Alpha-Omega.

Jean-Claude Lanoue, who was always anxious to tackle new technologies, wanted to design a new Omega System that used new, surface-mounted electronic components. Electronic devices usually have wires that go through holes in a printed circuit board where they are soldered to copper tracks. Those tracks are the wires that connect the components. The surface-mounted electronic components, as the name suggests, are connected on the same side of the circuit board on which they are mounted, which means that designers can use the other side of the board for other components. As a result, many more electronic parts can be packaged in a smaller space.

Many of our customers wanted smaller products that consumed less power but had more capability than the current systems, and so we tasked Jean-Claude to research this new technology and the new production techniques we would have to introduce. To make sure that Jean-Claude was never without work, we also tasked him to begin the design work on a new control and display unit that would have a color multiline display using a television-like cathode-ray tube.

The work that we wanted Jean-Claude to carry out was advanced research. It was an investment in our future. We did not plan to use those technologies in new products until we understood them properly.

That was typical of the way we worked. We shunned advanced research unless it had a particular product application. We did not have a separate research laboratory. Rather, we carried out all research in the product laboratories, which kept the research focused on a practical application.

Two Solitudes

A few years earlier the Canadian Marconi Avionics Division had set up a facility in Ottawa. Sol Rauch, the product champion for our very successful line of engine instruments, had opened the operation.

Sol and his small team of engineers first occupied one floor of a commercial building, where they pursued new product research and development and began to build expertise in systems engineering work. One of the company's long-range goals was to secure contracts with the Canadian government where Marconi would have the prime responsibility for buying and integrating various avionics for an entire airplane program.

Sol's group expanded, and a product development and production facility in the Ottawa area, to augment our main plant in Montreal, became a reality. The company bought land in the high-technology community of Kanata, west of Ottawa. Within a year, Sol and his team of engineers moved into the first unit of an expandable facility. By 1980 our Kanata facility was a growing concern. Other divisions of the company had plans to expand into the facility, but Sol Rauch and his team of avionics engineers occupied most of the space.

The latter part of the 1970s had been an unsettling period in Quebec. The separatist government, the Parti Quebecois, was the ruling party, and even after the defeat of the referendum on Quebec separation, many English-speaking Quebecors were leaving the province. Against that background, the company initiated the expansion into Ontario. Rumors about the impending closure of our Montreal operations became commonplace. How those rumors started I do not know. All we had done was move a handful of people from our Avionics Division and hired a few in Ottawa. However, the rumors persisted and became more frequent when the company bought land and started to build the facility in Kanata.

We devised many strategies to allay the fears of our Montreal employees and to gain the maximum benefit from our expanded capabilities. From the beginning, the goal of the company's Kanata operation was to handle product research and development. Work was not taken from the Montreal-based Avionics Division; new ideas came from both facilities, and we continued to manufacture all avionics in Montreal. It was only much later that the Kanata facility undertook some manufacturing operations.

Our decision to continue to concentrate manufacturing in Montreal allayed the fears of only some employees. Our engineers continued to suggest that the Kanata facility was the first step in a long-range secret plan to move research and development out of Montreal. Nothing could have been further from the truth.

Initially we limited the mandate for our Kanata colleagues to the preliminary design of new products, and the division's management committee devised a plan for transferring products from Kanata to Montreal, where we would complete their design. Now we faced a growing work culture difference between the two operations.

In Montreal, we tended to design products only when we were confident of their market potential. From the beginning of the design process, we employed concurrent engineering techniques, which was the practice of involving all product and production engineering disciplines early in the product design cycle.

Our colleagues in Kanata, because they focused on military markets, favored a different approach. Their technique was to produce, as quickly as possible, a prototype of a product that a potential customer could test. Then they hoped to secure a development contract from the customer to complete the product and ready it for production. The approach had some advantages but also a few disadvantages.

The Kanata engineers were very imaginative, and they came up with many novel product ideas. Our plan was to transfer a product only if a group in Montreal wanted the product. Engineers prefer to work on their own designs, not someone else's, and so we had to be sure that we transferred a prototype product only to a willing party.

Peter Gasser and I were the first managers to oversee the transfer of products from Kanata to Montreal. There were two that we thought we could sell into the civilian market. One was a small and effective fuel management system. The other was a device that monitored many parts of an airplane and displayed their condition on a unit in the cockpit. If we were successful with those transfers, we would be on our way to having a range of commercial products to build on the success of our Omega product line.

Elsewhere in the division, engineers were working on two other products that Peter and I thought would eventually be suitable for the commercial market. One, another project from our Kanata facility, was a new navigation system that would use a constellation of satellites. That U.S. Air Force sponsored navigation service is now known as the Global Positioning System, but at the time the civilian market was unsure about using it because of its military sponsorship. The situation was very similar to the one that we had faced initially when the U.S. Navy controlled the Omega System. Consequently, Sol's engineers were focusing their work on a design for military customers.

The other research that interested us was being done in Montreal. We had engineers conducting preliminary design work on a new aircraft landing system. Like the Omega System and the Global Positioning System, the landing system was an entirely new technology. However, unlike the other two, it had been started by the civilian aviation community. The Microwave Landing System, as the new system was called, was not due to be ready until 1995.

Ironically, both of the projects that Peter and I saw as significant to our long-range civilian avionics strategy relied on government-provided services. As events unfolded, both systems suffered similar types of delays and signal problems, as initially we had experienced with the Omega System.

Commercial Avionics Products

Early in January 1982 we reorganized ourselves once again. For me it was the fulfillment of a 10-year odyssey. We created a department called Commercial Avionics Products. Finally, this originally military company had fully recognized our commercial avionics business. Peter asked me to head the department. For 10 years the entire Omega Team had worked toward our own identity, and we had succeeded.

The Commercial Avionics Products department had two product groups: the Omega product line and the two products that we were transferring from our Kanata research and development operation. David Bailey succeeded me as the Omega product manager, and we promoted Tony Murfin to manage the two new products from Kanata. We called his department Flight Systems because pilots would use the two products, which he was about to inherit, to help them fly airplanes rather than navigate. Those two products were the Flight Advisory

Computer, which was a fuel management system, and the aircraft equipment Status Display System.

As our business and product base expanded, those departments became known as Commercial Navigation Systems and Commercial Flight Systems.

My forte was marketing and business development, and although Peter Gasser, David Bailey, and Tony Murfin had considerable talent in those areas, their forte was engineering. To take advantage of our respective strengths, we agreed that we did not have to follow a strict reporting hierarchy. David and Tony received guidance directly from Peter on technical matters that were beyond my capability, whereas I concentrated more on business management and market development. The arrangement worked well and shows, once again, how we worked as a team, without regard for our personal aspirations. We did not always agree—we were four different and strong personalities—but we did strive to make decisions that were in the best interests of our business, not ourselves.

We chose the name of our department carefully. The use of the word products, not markets, in Commercial Avionics Products was deliberate. Increasingly, military organizations were buying commercial products because, those products were less expensive than ones designed to military specifications. Our department's mandate was to produce commercial avionics but to sell them to the military, as well as to the commercial aviation sectors of the market. Initially, I frequently had to clarify our mandate with our own team, as well as the company's corporate management. We had a tendency to focus on commercial customers, and our corporate management would ask who was marketing to the military.

The marketing department for our Commercial Avionics Products had grown with our market. Henry Schlachta now had the responsibility for all of our North American commercial avionics activities. Tony Deveau had learned quickly since he joined the Omega team as a marketing trainee in 1978, and I put Tony in charge of our international marketing effort.

Two years earlier Bill Buker had joined our marketing group. Bill and I had gone to India together 10 years earlier, while I was still working for Marconi Avionics. Ever since, I had watched Bill progress as a technician and looked forward to the day when he would choose marketing as a career. His potential as a marketing person was obvious.

The general aviation market needed constant attention, with frequent visits to potential and existing customers and our fixed base operator dealers. Because most of those businesses were located in the United States, we hired two professionals who lived in the states.

David Hanley joined C.M.C. Electronics Inc., our U.S. subsidiary, in New Jersey. That company had grown out of the service center that we had set up to support Pan Am five years earlier. David's territory was the entire eastern region of the United States, and he covered business aircraft operators and the regional airlines.

Al Werner, who had been a pilot for one of our customers that operated a fleet of business aircraft lived and worked out of Cincinnati. He covered the central and western region of the United States. A few years later, Al left the company, and we hired Bill Finley, who lived in California. Bill Finley and Dave Hanley, seasoned marketing professionals, made a significant contribution to the company's commercial avionics business.

A new employee, Fred Prins, covered Europe, while Tony Deveau continued to sell in the market that he knew well: South America and all of the Pacific Rim countries.

We gave Carl Hanes the responsibility for marketing our new products. We knew we would have to continue to work closely with the regulatory agencies, especially as we launched new products, and Carl had special talents for that task.

For the first time in 10 years, I no longer had any frontline marketing responsibilities. I tried to manage the overall marketing effort, do my share of hands-on marketing, and coordinate the in-house activities of the two product lines. With more than 50 people to worry about, I found myself spending more time on personnel and overall business management matters. I did travel whenever I could because it was as important as ever that none of us lose contact where it matters most—the market.

As the manager of our newly formed Commercial Avionics Products department, I tried to streamline many of our internal processes that had consumed much time in the past.

Tony Sayegh joined our team as a contracts administrator in April, just four months after we formed the Commercial Avionics Products department. Tony had more experience than the job we offered him required, but to his credit he never complained and immediately worked long hours to help us with our contract, pricing, and proposal work. We needed the most help in those areas, and Tony made our entire contracts administration process more efficient. Within two years Tony had the opportunity to become the contracts manager for the division. Though Peter and I valued Tony's contribution to our team, we vigorously supported Tony's well-earned promotion.

I was so excited about the creation of the Commercial Avionics Products department that I had several marketing giveaways and posters made with Canadian Marconi Company Commercial Avionics Products emblazoned on them. I wanted to secure our identity both outside and inside the company. It was as if I was afraid that it would be taken away from us. Finally, it was too much for Lionel Leveille, who was now our vice president. He pointed out that this approach was an inefficient use of funds because the trinkets could not be used by other departments. He was right; we always had to be careful to make the most efficient use of our funds, and so I limited the number of giveaways that we made exclusively for our team.

David Bailey and his engineers had several new Omega products in development. The product line, from a technical standpoint, was under control, though its marketing remained head-to-head competitions.

In several respects, Tony Murfin had the more difficult task. He had to arrange for the transfer of two prototype products from Sol Rauch's team in Kanata. Afterward, he and his team had to complete their designs, ready them for production and help with their marketing. Under Tony's guidance and enthusiasm, his team of engineers performed wonders to bring those products to market. Like the rest of our team, those engineers showed a considerable amount of dedication and skill to accomplish what was, at times, an extremely difficult task.

Inheriting the products and getting them ready for production turned out to be more difficult tasks than we had imagined. The Status Display System monitored more than 100 parts of an airplane, including several parts of the engine and the fuel system. Normally there are many red and amber lamps in an airplane's cockpit, as well as gauges, that alert a pilot to a particular problem. The cockpit unit of our Status Display System presented all that information on one display. It was a brilliant idea.

The Flight Advisory Computer was a gem of a product. After a pilot had entered information about a planned flight, the system would calculate the most efficient ascent, cruise, and descent speeds and cruise altitude. The system provided that and other information so that a pilot would know what to do to save fuel. All of that was done with electronics within one small cockpit mounted unit.

The Flight Advisory Computer was brilliant in concept, and it came when many airlines were looking for devices to conserve fuel because its cost was rising. A U.S. avionics firm, working with Boeing, was designing a more elaborate performance management system that would succeed the one already in use on the Boeing B-727s. Rockwell-Collins, Sperry, and other companies in the United States and Europe were all working on similar systems. Some were comprehensive flight management systems that included navigation, communications, and airplane performance management. Others, like our Flight Advisory Computer, performed just the fuel management task. Depending upon the operations they performed, those products ranged in price from $15,000 to $80,000. Our Flight Advisory Computer was at the low end of the price range.

Most of our competitors' equipment had a cathode-ray tube display with a keyboard, which was mounted in the cockpit, plus one or more boxes of electronics. Ours had everything in one unit and was about the size of a car radio.

Once, when I was showing our Flight Advisory Computer to a group of Boeing engineers in Seattle, they were incredulous that we had put so many features in such a small package. All the other comparable equipment they had seen had two, or more, larger units. To prove I was not hiding anything, they asked me to remove the covers from the only other piece of equipment that I had in the conference room. It was a power converter so that I could use a standard power outlet to run the equipment. They were still skeptical. I was beginning to learn how an inappropriately packaged product can give a prospective customer the wrong impression. That product was comparable to those selling for $50,000, but it looked like, and was, a product that sold for around $20,000, and so it was treated accordingly.

That visit to Boeing occurred a year after Tony and his team of engineers had begun work on their newly inherited products. Soon after they started work, they found ways to make the product easier to produce. Immediately, we faced a problem that occurs when companies transfer products from one engineering group to another.

Several people accused Tony and his engineers of needlessly redesigning the product to satisfy their own creative needs. That was not the case, and we had to hold many meetings to justify our intentions before our management would sanction the redesign work.

We faced another problem because of our lack of experience with those types of products. Although we had produced many versions of our Omega products, they were not unique to a particular type of airplane. A CMA-771 that we had first installed in a Canadair Challenger business jet also would work in a Boeing B-737. That was not the case with either the Status Display System or the Flight Advisory Computer. Both products had to have a special set of electronics and software to work in a particular aircraft. The Flight Advisory Computer had to have in its memory the unique performance characteristics of the airplane, so that it could calculate the best flight profile to save fuel.

After talking to potential customers, we began work to customize the Status Display System for the Canadair Challenger and Gulfstream business jet aircraft. No airlines or airframe manufacturers showed an interest in the Status Display System. However, the airlines were interested in saving fuel. Tony's team worked on versions of the Flight Advisory Computer that would function in B-737 and B-727 airplanes because the operators of those airplanes were our most likely customers.

Those products were only moderately successful. If we had not been in such a rush but had listened more closely, we would have waited until an airframe manufacturer or a large established airline had chosen the Flight Advisory Computer. Then we would have changed the design to meet their exact needs. The outcome of this venture, though, did not become apparent to us for several more years. When we first inherited the products, in 1982, we were enthusiastic, energetic, and anxious to add more commercial products to complement our extremely successful line of Omega Navigation Systems.

New Markets and Products

On May 3, 1982, I called Christina to wish her a happy birthday. I remember the call because it was her 40th birthday, and once again business had prevented me from being present at an important family event. I also remember the call because I made it from China.

A few weeks earlier we had received a telex from John Che, the Canadian commercial counsellor in China, asking us to send a delegation to talk to the Chinese about our Omega and telex exchange equipment. The same day that we received John Che's telex, we got a call from Igor Neudachin, our representative for the Soviet Union. Igor told us that Aeroflot was planning a major purchase of Omega Navigation Systems.

Igor was a huge bear of a man, but as so often is the case with large men, he was gentle and, most of the time, spoke softly. Although he was a Russian, he had lived in England for most of his life. In the evenings, with a vodka or cognac in hand, Igor enthralled us with stories about his miraculous escape from Singapore during World War II. His father was a Russian diplomat, but we never heard what happened to his parents. As Igor told it, at the age of seven, he escaped alone on the last evacuation boat. Igor somehow, which was also never quite clear, ended up in England.

Regardless, Igor had ready access to the Soviet Union. While most non-Soviets had to wait weeks, or months, for a visa, Igor was able to enter and exit Russia at will. Igor had so many contacts that we sometimes wondered about the independence of his business. We jokingly, and openly, referred to him as our KGB agent. When he died several years ago, we mourned the loss of a friend.

Igor conducted his business dealings with us with utmost integrity. During one particularly lengthy negotiation session with Aeroflot and Aviaexport, the Soviet Union's procurement agency, we reached an impasse. Igor spoke separately with the Soviet negotiating team and then told us that if we reduced our price by a further 4 percent, they would sign a contract. Already they had negotiated a very competitive price, and we did not have any more money to give away. Igor provided the solution: if we reduced our price by 2 percent, he would reduce the commission that we had agreed to pay him by 2 percent. That night we celebrated the contract with our Russian guests at a restaurant in Montreal with the usual copious quantities of vodka.

That event occurred 18 months after the point I left our story. Peter Gasser and I discussed who should go to Russia and who to China. Although I had been to Russia twice before and knew what to expect, I was anxious to visit China for the first time. We flipped a coin. I went to China, while Peter went to Russia.

If the size of our delegation surprised the Chinese, they were too polite to say anything. The first Canadian Marconi delegation to China consisted of three people. Don Mactaggart and Paul Caden went to talk about our telex exchange, and I was the Omega expert.

We did not have clear instructions to help us prepare for the trip. All we knew was that we would have meetings in Beijing and Shanghai. John Che tried very hard, but to no avail, to find out exactly what the Chinese wanted us to do once we arrived. We knew that we would have to make presentations, but we did not know if those sessions would be 30-minute sales talks or 4-hour seminars.

To safeguard myself for any situation, I took with me the 300 slides on the Omega System and our equipment that I had collected during the last 10 years.

Running had become a serious hobby for me, and I was then running about 50 miles a week, training for my third marathon. I knew very little about China, and what I did know led me to think that the Chinese would restrict our movements. Because I wanted to run while we were in China, I asked our embassy colleagues if I would require special permission to run in the streets. Their answer was that I was free to go wherever I pleased and that our military attaché, also a runner, would run with me one morning to show me around.

They must have thought that I was rather ignorant or naive to be asking such questions. I expected the restrictions that we had sensed in the Soviet Union and the poverty that had saddened me in India. I was wrong on both accounts.

Soon after our arrival, I was glad I had brought 300 slides covering everything that I knew, and some things that I did not know, about the Omega System. At an introductory meeting with our official guides and hosts, the Chinese told us that they wanted us to provide two three-day seminars. Our audience, they explained, would include engineers, researchers, customers, and others. The others, we assumed, would be there to make sure that neither we nor the Chinese strayed from the prescribed program or topics. They told us that none of the seminar attendees spoke English, and so they would provide a translator.

We spent a day with our assigned translators going over the material and technical terms that we would be using. The next day we were taken to our respective seminar locations. Don and Paul went one way, and I went another. I was on my own with some 30 Chinese and a large tea cup that a Chinese person regularly refilled.

Our official hosts were gracious and hospitable, and we enjoyed several typical Chinese banquets. Around a large circular Chinese banquet table, we enjoyed good companionship and a multitude of Chinese delicacies. We learned Chinese table etiquette, where it is impolite to serve yourself; your neighbor serves you, and you serve him or her. Our hosts, like those attending our seminars, and the general population in the streets were warm, friendly, and cheerful.

On the first morning after our arrival, I was up early and anxious to go running. The Canadian military attaché met me outside the hotel, and we proceeded to explore Beijing. The hotel was a mile or two east of Tiananmen Square on the main east-west boulevard. The Imperial Palace—the Forbidden City—borders that huge square to the north. To the west is the Great Hall of the People, and opposite the Great Hall on the east side of the square are the Museums of the Chinese Revolution and History. At the south end of the square is the Chairman Mao Memorial Hall. All are magnificent buildings.

From the square we jogged through Beilhai Park, which is next to and extends north and south along the entire length of the Imperial Palace. Then we headed back eastward, winding our way through residential areas. Most of the homes were small, single-story adobe-like dwellings clustered around small courtyards. The bricks and cement appeared to be both old and of poor quality, though I expect that had more to do with the type of clay in the region than workmanship.

I remember having the impression that the Chinese lived a harsh and frugal life, but they were well dressed, friendly, and cheerful. I did not see any of the poverty or undernourishment that I had seen in India. Most of the men wore well-tailored high-neck suits, of the same pattern, in either grey or black. In the West we refer to those as Mao suits. By contrast, many of the young women, and particularly the young children, were dressed in bright clothes. We passed the enormous Beijing Workers Gymnasium and Stadium. No wonder we did not see any Chinese running; they were probably all in the enclosed gymnasium.

What we did see, and had to be constantly aware of, were thousands of bicycles. After a few mornings running around Beijing, I noticed that the Chinese were uniform even in their bicycling habits. I was overtaking them, and because I knew how fast I was running, I estimated that they cycle at a constant speed of 7 miles per hour. My quicker pace also prevented the first row of cyclists, riding eight abreast in a special bicycle lane, from running me down. Over my shoulder I could see hundreds of rows of cyclists, all relentlessly moving forward.

I wondered how the Chinese picked out their bicycles at the bike stand after work. The hundreds of bicycles all looked the same: black, with upright curved handlebars, mud guards, and no gears, the type of bicycle that I remember my father riding in the 1950s. I asked one of our hosts how they locate their bikes. "Oh, that's easy. We tie a small colored ribbon or other item to the bike, so we know which one is ours," he replied. Then he added, "It also helps to remember exactly which slot we left it in."

On other mornings I ran alone, exploring different parts of Beijing, including the grounds of the Forbidden City. There, I ran slowly so that I could marvel at the intricate architecture, statues, sundials, and other artifacts. Occasionally, I would stop to watch a group of Chinese performing the exacting art of tai-chi, one of their most popular forms of exercise, which requires considerable concentration.

When we completed our seminars in Beijing, we had two days of sightseeing before leaving for Shanghai. Naturally, we wanted to see the Great Wall of China, and so our hosts arranged a car and driver for us to visit the Badaling section of the wall. That part of the wall, which is about a two-hour drive from Beijing, is the most photographed section; it is the part of the wall that we see in our Western newspapers during a state visit.

The Chinese built the Great Wall of China, the world's longest wall fortification, to protect them from their Mongolian neighbors to the north. The wall is approximately 1400 miles long, 25 feet high, and 12 feet wide. Construction of the wall started during the Ch'in dynasty, in 246 B.C., though most of it was built during the Ming dynasty, more than 1500 years later. As we walked on the wall, I could not help thinking that we were walking on the tomb of thousands of people who had died building that fortification. Like every visitor to the Great

Wall, we were awestruck by its magnificence and the sight of it winding its way up and over rolling hills, some of which were quite steep.

Almost as impressive as the wall were the Ming tombs, which were en route to Badaling, and in particular the boulevard leading to them. The boulevard was lined with magnificent life-sized stone carvings of animals.

For our last night in Beijing, we went to a restaurant that reportedly served the best Peking duck in Beijing, and that was all it served. The restaurant was in the heart of the business district on the second floor of a nondescript building, the restaurant was equally ordinary and spartan. The reports were correct because any sense of ordinariness went away when we took our first bite of its Peking duck.

The next day we flew to the bustling seaport city of Shanghai. There, we once again held three-day seminars. Paul and Don toured a telecommunications facility and met with its engineers. I held forth in a large conference room in our hotel, which was located about 3 miles from the center of Shanghai.

Determined to see as much as possible of Shanghai, I woke early each morning to go running. On the first two days I ran from the hotel to the center of Shanghai and then down the main street to the waterfront. I only had to remember to make one turn on that route. On the third morning, I decided to be more adventurous and to investigate some side streets, but to avoid getting lost, I periodically returned to the main street.

For several blocks I successfully negotiated the side streets and returned to cross over the main street. I was enjoying the sights, sounds, and smells of Shanghai and preparing for another working day. Several people shouted, "Hello, Mr. Canada." I always wore a tee shirt with "Canada" printed on the front. The Chinese were anxious to practice their English and to be friendly. I turned to head back once more to the main street, but I had forgotten that it made a sharp bend in that area, and so, when I returned, I missed my only point of reference. After 30 minutes of frantic running, I had to admit to myself that I was lost.

I confess to mild panic. I could not read the street signs, and not one of the several people whom I stopped spoke enough English to help me. Even taxi stand signs were unintelligible. Being lost in Shanghai, when you cannot read or speak Chinese, is a very lonely experience.

I ran in a state of panic for another 15 minutes, and then ahead I saw the backs of two other joggers, the only other joggers I had seen all week. When I caught up with them, I was overcome with relief to discover they were Americans. Fortunately, they were not lost, and so I ran with them to their hotel, some five miles distant, where I took a taxi back to my hotel. It is degrading for a runner to return in a taxi, but I was too relieved to care what anyone thought.

My seminar translator was entering the hotel as I arrived in the taxi. I explained to him what had happened, and asked him to convey my apologies to the group, and said I would be 15 minutes late. I rushed to my room, showered, changed, and rushed back downstairs to the conference room. My hosts greeted me with a warm applause and many smiling faces. While polite and sympathetic, my Chinese audience was also amused with my early morning adventure.

The following Saturday, I went for a 20-mile run straight down the main street and then along the waterfront. The sights that I saw during that run were those of an industrialized city, as well as a major seaport. Shanghai is home to more than 13 million Chinese, half of whom said "good morning" to me.

Our trip to China was a wonderful experience, and I hoped that business would result, though I had several concerns. During conversations with officials in China, I was asked if Canadian Marconi would allow them to manufacture our product.

Under some circumstances that might have been possible, but they also wanted to build equipment for our customers outside of China. We had an obligation to our own people and our factory, and our market was not large enough to warrant two sources of manufacture. Also, the Chinese wanted us to pay them standard North American labor rates, not their actual rates. I could see that doing business in China was going to require a considerable investment in resources.

With some reservations, we agreed to support a flight test program in China. The flight tests occurred about one year after my visit. We had to send field engineers and instructors to help with the installation and to train their flight crews. The business, though, did not follow.

Peter Gasser had a more successful trip to Moscow. He also had to agree to a flight test program with Aeroflot, and within 18 months we signed the first of several large contracts with the airline. The contract negotiations, while strenuous, were easier on our constitutions than the evenings we spent entertaining our guests when they were in Montreal. Fortunately, when we tired, Igor was always prepared to keep the party going, while we went home to bed.

To celebrate the signing of our first contract, we treated our guests to a dinner and cabaret at one of Montreal's better nightclubs. The show included a comedian who also picked pockets, and we narrowly avoided an international incident. He visited with the dinner guests and then returned to the stage, where he revealed the items he had collected. With appropriate humor, he returned watches, bracelets, belts, and other items to their surprised owners.

The comedian had one wallet left, which he opened. On finding the identification of one of our Soviet guests, he proceeded to make jokes about a KGB agent in the audience. It was a tense moment. Our guest was embarrassed but played along. The lesson we learned that night was not to put customers in a potentially embarrassing situation. Whenever we returned to the nightclub with other customers, we either checked the program first or warned our guests.

On May 30, 1982, two weeks after returning from China, I ran the Montreal marathon. It was nearly my last one. The temperature was a blistering 30 degrees Celsius with little shade on the route. At 30 kilometers I was ready to give up, as had many others already. Christina pleaded with me to stop before I did myself real mischief, but John Rogers, knowing the challenge of the marathon, was there, providing encouragement. He was not running in Montreal that year because he had run the Boston Marathon a few weeks earlier. John joined me for the last 12 kilometers, and with his help, I made it to the finish line. With less than 100 meters to go, some runners were sitting on the curbside with nothing left. Such was the challenge of the marathon. I recorded a personal worst time of 3 hours and 52 minutes.

Of our new commercial avionics products, the Flight Advisory Computer was our biggest challenge. Our engineers had to learn about airplane performance, an aspect of flight that they did not have to consider when designing navigation systems. Also, we needed a launch customer, preferably an airframe manufacturer.

One result of President Reagan's deregulation of the U.S. airline industry was the emergence of several new carriers. Some were small, regional carriers that provided a feeder service between small towns and the airports serviced by the large carriers. Others were new domestic airlines in their own right. One such new airline was People Express, which began operations out of Newark Airport, New York's third airport.

People Express was an employee-owned, no-frills airline. Its airplanes, mostly secondhand Boeing B-727s and B-737s, were comfortable, clean, and tastefully painted, and their service was attentive but minimal. People Express offered low-cost air travel, initially in the northeastern part of the United States, and then later throughout the eastern seaboard and across the Atlantic.

We targeted Boeing and People Express as the most likely launch customers for the Flight Advisory Computer. Boeing was not particularly enthusiastic because it had a similar program under way with another supplier, but the senior managers at People Express showed considerable interest. Consequently, Tony's team concentrated on producing the software that the Flight Advisory Computer needed to work in the several versions of B-727s and B-737s that People Express operated.

David Hanley, Henry Schlachta, Tony Murfin, and I made frequent visits to People Express to obtain information about its airplanes and to convince it to buy the Flight Advisory Computer and the CMA-771 Alpha-Omega Navigation System. Both products were ideal for the type of routes that People Express flew.

Because several companies were now manufacturing flight management systems, we accelerated our Omega sensor design work. We recognized that, for companies such as ours, our business environment was changing. More often our prospective customers were companies that were providing the larger integrated systems that our mutual customers now required. If we had not recognized that change, we might have failed, but once again we had the good sense to be flexible.

Our product development and marketing strategies evolved, as our commercial product base grew and the market changed. During the early 1980s, we concentrated on five major strategies.

We designed variants of our Omega Navigation Systems that would work with flight management systems. The five principal manufacturers of those systems were our target market.

To ensure that we remained competitive and responsive to customers who wanted to buy a stand-alone navigation system, we continued to improve the CMA-771 Alpha-Omega.

Also, we designed a low-cost control and display unit to work with our Omega sensors. With the addition of a control unit, we had a low-cost navigation system to offer to operators of small airplanes and helicopters. That way we made full use of, and expanded the market for, equipment that we had originally designed as a sensor.

We targeted the Status Display System for the Canadair Challenger and Gulfstream business jets and the Flight Advisory Computer for Boeing and People Express for B-727 and B-737 airplanes.

Our colleagues in Kanata were developing an Engine Health Monitor for military airplanes, and because the airlines had expressed a need for such a product, I explored the possibility of a joint venture with Boeing. I knew from our early experiences with the Flight Advisory Computer that we should not design another product that was unique to a particular airplane type unless it was as a joint venture with the airframe manufacturer. Rather than transfer the product, we assigned an engineer to the Kanata team to learn about the Engine Health Monitor, so that we could talk intelligently to prospective customers.

We had several ideas for products in the future, but I wanted to make sure that they fitted a logical sequence. We wanted to reach a situation where we offered a range of related avionics products.

The Microwave Landing System was one product that had promise. It also fitted well with our navigation and fuel management systems. A civilian version of the military satellite navigation system, which Sol's team in Kanata was designing, also would be an obvious addition to our range of commercial avionics.

Two new avionics requirements that were emerging captured my attention. The FAA was promoting the design of a collision detection and avoidance system because there continued to be near misses between aircraft. Midair collisions were extremely rare, and the FAA wanted to keep it that way. The airlines wanted an air-to-ground digital data link so that they could exchange flight information with the ground electronically and not by voice communication. There was talk of a communications satellite system to support that data link.

We had no shortage of ideas for new products and markets. However, ideas do not cost money; only their implementation does. Our human and financial resources remained limited, and so we embarked on a period of careful long-range planning. We spoke to our customers and potential customers. We attended various industry meetings that addressed future requirements, and we funded some small-scale new product research studies. It was not until early in 1984 that I was comfortable putting together a strategic business plan for our commercial avionics business. By then, Peter Gasser was vice president of our Commercial Communications Division, and I was the manager of Commercial Avionics Products and the division's product support groups. Those promotions had occurred in March of 1983, and it was the first time in 10 years that Peter and I had not worked closely together on the same team.

Tony Deveau, who had been with us for several years, had the opportunity to expand his career in our military products group, and so he moved. I put Henry Schlachta in charge of all our commercial avionics marketing, and Eric Ford joined us from another department.

During the years of our transition into the civilian market, we had been fortunate to have several dedicated marketing professionals on our team. It was always a small team, but everyone worked hard to achieve our goals. Eric brought new insights and experience to our team, and like most others who joined us during our transition, he became a valuable contributor. He was also good company and had a remarkable sense of humor, always important in a frenetic working environment.

Product Improvement Initiatives

At the International Omega Association's annual meeting in 1985, I presented a paper titled "Omega Cost Control and Product Improvement Initiatives—A Manufacturer's View." In the paper I described the technologies and productivity improvement techniques we had employed to improve our Omega products while substantially reducing their costs.

In 1972 our first Omega product, the CMA-719, had a list price of $80,000. If we had done nothing but accept the escalation that was occurring for such products, by 1985 the CMA-719 would have cost $219,000. Instead, we were offering the far superior CMA-771 Alpha-Omega Navigation System at a list price of $55,000.

By now we had developed a low-cost control and display unit for our small CMA-734 Omega Navigation System, and we had made productivity improvements in the main electronics unit. As a result, we could sell the CMA-734 Arrow, as we called the new version, at a price that was 40 percent less than that of the first version sold in 1976.

We had even managed to contain the cost increases of our first airline systems, the CMA-740 and CMA-771, which we introduced in 1977 and 1979, respectively. By 1985 we had increased their prices by only 15 percent and not the 60 percent that labor and material costs had increased during the intervening eight years.

In my paper I referred to this continuous product improvement process as an evolutionary approach to cost control and product competitiveness. Simply stated, our Omega products evolved as customer requirements changed and new technologies became available. We used new technologies, though, only if they improved our product for our customers or its manufacturing process.

Electronic component manufacturers made several significant improvements in their products during the early years of the 1980s. The most significant for us was their introduction of surface-mounted devices. The surface-mounted electronic components were much smaller and more powerful than their predecessors. Also, by using those devices, we could now mount components on both sides of the printed circuit board.

Our engineers had little difficulty designing circuits using those devices; the big challenge was on the production line. There our production staff had to use entirely new techniques and equipment. Jean-Claude Lanoue, the Omega team's technology expert, led our design and production planning effort for that new technology. He was so successful that we were one of the few companies that did not have significant problems introducing those new electronic devices.

We first used those components in an Omega product that we introduced in 1986. For the first time in 10 years we could not offer our customers an upgrade of their existing systems. However, we did continue to improve those products with the new technologies. We first introduced our new Omega product, the CMA-764, as an Omega sensor to work with flight management systems. Later, we added a control and display unit so that it could work as a stand-alone navigation system. Still later, it evolved into a full navigation management system, including both an Omega System and satellite navigation capability.

Successful companies pay continuous attention to the costs of their products. In my International Omega Association paper, I wrote the following: "The evolutionary approach to cost control requires the maintenance of a cohesive engineering team, who are constantly reviewing and implementing new concepts and technologies. It requires the maintenance of production methods, quality assurance, and integrated logistics support teams, who are constantly seeking for better and more efficient methods for building and supporting the product."

In the same paper I outlined another, and often overlooked, method by which we remained cost competitive. As products evolve, their cost often also grows. Sometimes that is unavoidable; more often, though, companies become lazy. By using a planned approach to cost control, companies can avoid such cost increases. Sometimes that is called the cost scrub. Whatever we call the method, it is usually a focused activity, performed periodically, to find ways to reduce the cost of a product without jeopardizing its performance.

We performed several such cost reduction exercises. Once, when I was the Omega product line manager, I commandeered a conference room for two weeks. On the conference room table we set out assembled and disassembled Omega products. I then invited anyone from throughout our organization to go, look, touch, fiddle, and offer their suggestions for cost reductions. As an incentive, we offered dinner for two in a fine Montreal restaurant to the person who had the best idea. After two weeks, we awarded two dinners for two. Out of that exercise we had six ideas that saved 10 percent of our production cost. On other occasions we adopted a more conventional approach. We met with the experts of the various disciplines involved in designing, producing, and supporting the product and asked them to justify their costs. We solicited their suggestions. As managers we understood that those people working directly with a product are the ones most likely to know how to make it better. That approach also gave everyone a feeling of involvement. Today that approach has a name: Total Quality Management.

Through these and other initiatives, we maintained our momentum in the Omega market. We produced three distinct generations of our CMA-771 airline Omega Navigation System and six of the smaller CMA-734 system. Ten years after we introduced our first airline system, we offered a comparable product that was one-third the size and less expensive.

By 1986 we were well on our way to having several other commercial products for airlines, for general aviation companies, and for military applications that did not need a ruggedized military product. Our transition to the civilian market was almost complete.

I have already quoted from a paper that I wrote in 1985. I end this chapter with another quote, from the same paper, to show that we understood and practiced good business habits and learned from experience and common sense, not from current management books: "It is not sufficient for a manufacturer to push only the state-of-the-art of a product. One must also push the process for developing, manufacturing, and supporting the product in order to remain competitive. Oftentimes this requires a heavy injection of capital investment as well as employee retraining." I should have included the fact that a manufacturer must also listen and respond to a customer's real, not perceived, requirements.

We did all of that and more, which is why during 10 years we sold more than 2600 Omega Navigation Systems. More importantly, during that time 56 of the world's airlines in 45 countries began using our products. Now we had market momentum, but we understood that we had to continue to earn the right to that market. Our past successes would help us to maintain our future market, but they would not guarantee our future success.

18

Putting It All Together

*The thing always happens that you really believe in;
and the belief in a thing makes it happen.*
—Frank Lloyd Wright

Canadian Marconi set up its Avionics Division after it helped to pioneer navigation equipment that used the Doppler effect. Aircraft navigation was and remains the division's heritage. It is a part of the avionics business in which the company continues to excel. Naturally, when we were considering new products, we thought about navigation or navigation-related products. We planned our product development to give us a logical mix of products—especially our commercial avionics. That is one reason why we designed a Microwave Landing System receiver. The Microwave Landing System, like the Omega System and the Global Positioning System before it, has been adopted by the world's aviation community.

We began this venture with our sights set on the airlines; however, our first opportunity was with the U.S. Air Force. I have singled out a story about this product in a book that has emphasized the Omega System because it shows how we applied all the lessons that we had worked so hard to learn and remember.

This chapter contains all the drama that usually accompanies a government procurement competition. There were delays as Congress questioned some Air Force plans and budgets and other delays as the ground rules for the competition changed. We began the campaign in 1984, and it lasted four years. We put together sound engineering and marketing plans, followed them, and then changed them to meet new circumstances. We started early; we helped the Air Force to develop several parts of its specifications, and those we did not help to write we made sure we understood. We analyzed our competitors and all the factors that might jeopardize our success. In short, we did our homework.

We pursued our campaign relentlessly. For four years we accepted late nights, deadlines, disappointments, joy, and lost Christmas and lost summer vacations, but we always worked to win.

A New Landing System

For more than 40 years the Instrument Landing System has guided airplanes during the final approach to a runway. The Instrument Landing System has some limitations, and so as air traffic increased, the aviation community began the search for a new system. By 1967 those leading the search had more than 50 different systems to consider. The world's civilian and military communities wanted a single system that they all could use.

Those studies came to a conclusion in 1978 when, after much heated debate about the competing techniques, the International Civil Aviation Authority adopted the Microwave Landing System.

The ground stations of both the Instrument Landing System and the Microwave Landing System aim an electronic beam on the approach path to a runway, but the Microwave Landing System beam is much larger. The broader coverage means that several aircraft can lock on to the Microwave Landing System, which allows air traffic controllers to schedule the airplanes closer together on the approach path. When the Microwave Landing System is fully operational, we will no longer have to spend time circling airports, waiting for our turn to land.*

The world's civilian aviation authorities agreed to begin installing Microwave Landing Systems in 1990 and to have all airports equipped by the year 2000. NATO and several other military organizations also have adopted the system.

An additional benefit of this new landing system is that its ground equipment is easy to install and can be portable. That feature particularly interested military organizations, which often have to establish an airstrip in a remote location.

In 1976 the U.S. Army, wanting to take advantage of that feature, put a line item in its budget to fund the development of a portable Microwave Landing System. The U.S. Congress rejected the request and told the Army to use standard civilian equipment. It was a theme that Congress would repeat 10 years later.

Canadian Marconi, always on the lookout for new products, assigned John Haberl to participate in several of the early committees investigating techniques for the new landing system. In 1982, with that background knowledge, the company decided to design a Microwave Landing System receiver. Dave Allcock

---

*The Microwave Landing System did not become fully operational for a variety of reasons; one reason was the emergence of the U.S. Global Positioning System.

led a small team of engineers, who concentrated first on the most complex part, the microwave front-end of the receiver.

In 1983 the U.S. Department of Defense (DOD) decided to start a program that would lead to the procurement of Microwave Landing System ground stations and aircraft receivers for the Air Force, Navy, and Army. The U.S. Navy was already using a similar system on some ships, but Pentagon officials selected the Air Force to coordinate the triservice procurement program.

In August, the Air Force released a Microwave Landing System implementation plan. The plan recognized the commercial design activities that were then under way, but it also addressed the special needs of the three services. The Air Force identified requirements for three types of receivers.

One was a commercial receiver with changes to meet some special needs of military transport airplanes. The Air Force decided that its fighter aircraft would require a specially designed military receiver. Aircraft landing on a rolling ship present a unique problem, and so the plan listed a third receiver for use in Navy shipborne aircraft.

The implementation plan included estimated requirements for 5000 modified commercial receivers and 7500 special military receivers. We had started the design of a receiver, and so here was a market opportunity that we could not ignore. The implementation plan also included requirements for commercial ground stations and several special military versions.

Early in 1984 the FAA awarded the Hazeltine Corporation the first in an expected series of contracts for Microwave Landing System ground stations to be used at civilian airports. Even before deliveries to the FAA, Hazeltine, the Bendix Corporation, and Wilcox Electric sold a few other stations to municipal airport authorities. Those installations and some experimental ones gave those of us who were designing receivers an opportunity to test our systems before the full-scale changeover from the Instrument Landing System started.

The Campaign Begins

In 1984 we transferred the Microwave Landing System receiver design work to our Commercial Avionics Group. I had been managing that group for one year since Peter Gasser became vice president of another Canadian Marconi division.

The group had two departments, Commercial Navigation Systems and Commercial Flight Systems. The Commercial Navigation Systems department, under the capable leadership of David Bailey, handled all our Omega products. The Omega product line was an unqualified success, and David was managing the design of new Omega products to ensure the product line's continued success. The other was our Commercial Flight Systems department, under the stewardship of Tony Murfin. Tony was managing the completion of the fuel management system and the product that monitored and displayed the state of many aircraft systems. His team had done a remarkable job to make those products ready for production.

Tony, Carl Hanes, and others were also very active in the market trying to secure customers for both products. It was a very difficult task, but we were confident that the new airline People Express would soon become our launch customer for the fuel management system, and we expected the airframe manufacturer Canadair to buy the Status Display System. Though we still had a lot of work to do in the laboratory and the market, we were optimistic about the future for both products.

Tony had shown a remarkable amount of energy, and so I was not surprised as we discussed transferring the Microwave Landing System receiver work, when he asked to take on that project as an additional responsibility. We granted him his wish. Tony immediately tackled his additional assignment with the same amount of energy and enthusiasm he had shown in the past. Sometimes I found it difficult to keep up with Tony, especially his understandable impatience for more engineers. I could not approve all his requests for more staff, and so like all our other departments, Tony's remained a lean operation. I never ceased to marvel at the accomplishments of that team. Their work was rarely easy, and the product's design and its market became two of our biggest challenges.

In January 1984 our U.S. company, C.M.C. Electronics, reported that DOD was preparing to buy Microwave Landing Systems. We visited the Canadian Department of National Defense in January and February to discuss its requirements, and a possible flight test program. We wanted to be ready for the first question when we visited the U.S. Air Force: "What is your own government doing?"

The major airlines were not rushing to buy Microwave Landing Systems. As usual, they wanted more proof of the benefits of the system.

In the mid-1980s everyone expected the business jet operators and regional airlines to be the first users, and so that is where we focused our civilian market sales effort. In April 1984, Carl Hanes presented a paper on our Microwave Landing System receiver design work at the Aircraft Electronics Association convention. That convention attracted the operators of business jet airplanes and the smaller domestic airlines.

As we had done, when we helped to pioneer the Omega System, we used every opportunity to promote the Microwave Landing System—we were preparing our market.

Also in April, C.M.C. Electronics provided us with an update on the DOD procurement program. The Air Force expected to issue Request for Proposals for commercial receivers in June. We were already too late. We had not even visited the Air Force. In three months we could not hope to understand properly the customer's requirements and our competition or to conduct the comprehensive marketing campaign that is required to win.

The prospect of selling so many receivers, and so soon after we had started our design work, was an enticing prospect. We could not ignore it, no matter how slim our chances of success. The Air Force wanted receivers beginning in 1987 or 1988. We could accommodate that schedule because we expected to have our receiver design finished within two years. The Air Force planned to issue a second Request for Proposals later in 1984 for special military Microwave Landing System receivers. The military market was developing more quickly than we had expected.

Both the Bendix Corporation and the Sperry Corporation had designed Microwave Landing System receivers, and so those corporations would be our competitors in the Air Force competition. Clearly, they had an advantage over Canadian Marconi. They had products that were already flying, while our product was still on our engineering benches.

The Bendix and Sperry receivers provided the minimum information required for an airplane to use the Microwave Landing System during its approach to a runway. Although our receiver would be more expensive, we were designing it with many more features so that users could take advantage of all the capabilities of the system.

Because the landing system was new and not yet in operation, the aviation community was still debating how it would use all of the system's features. That meant we had to both guess what the procedures might be and, once again, work with the regulatory agencies, helping them to develop the procedures. Carl Hanes, of our marketing department, working with Tony Murfin, carried out that job for several years.

Soon after we received the April report from our U.S. company, the Air Force announced it would not release the Request for Proposals in June 1984. Perhaps we would have enough time after all.

We were pleased about the delay for other reasons. For some time we had been discussing a joint development program with our colleagues at Marconi Italiana. The Italian Air Force was about to start a Microwave Landing System receiver competition for its Tornado fighter aircraft. That competition would be important because the winner would probably also provide receivers for the Tornado fleets of other European countries. Canadian Marconi had the Microwave Landing System receiver technology, and Marconi Italiana had the political advantage. We both knew that we could not win that competition without significant European content.

As usual with the Italians, everything had to be done at breakneck speed, and so for three months during the summer, while we were attending to the U.S. Air Force, we exchanged visits with our colleagues at Marconi Italiana in Genoa.

By September, we had agreed on the basic terms for a joint development program, and together we had responded to a complex Request for a Proposal. Despite the initial frenetic activity, a year passed before the Italians announced the outcome of the competition. We won the first major Microwave Landing System receiver contract, which meant more work for Tony and his already overworked team.

On June 13, 1984, Sohel Fares and I visited Wright–Patterson Air Force Base, the home of the Air Force Avionics Systems Division, which we expected to lead the triservice Microwave Landing System receiver procurement program. Sohel was Tony Murfin's lead engineer for our receiver design work.

Realizing that a company's best marketing people are its engineers, we involved ours frequently in marketing activities. Engineers have the knowledge to assess a customer's requirements properly and to help the customer to produce realistic specifications. They are also able to commit to designing a product to those specifications. After all, they are the ones who will have to struggle on the bench to make those commitments a reality. Marketing people have the skills, interest, and knowledge to help execute a winning marketing strategy, but they must use other skilled professionals—the entire team.

During our visit we heard that the Air Force had made many changes to its Microwave Landing System procurement program. One change was that its aircraft operators now wanted four, not three, different types of receivers. We were pleased to hear that most of the receivers would still be commercial receivers with some special additional features.

The schedule they outlined by the Air Force was more like a usual military procurement program than the one we had heard a few months earlier. The Avionics Systems Division planned to pay a small consulting company to study the military requirements in detail and to recommend a plan of action. The division expected the study to take a year, and then it would release the Request for Proposals in July 1986. It would take two more years to obtain Pentagon and congressional approval for the program, to write the specifications, and to issue the Request for Proposals.

One piece of information that our colleagues at Wright–Patterson Air Force Base passed to us made us realize their program may not be exactly as they described. That was not unusual, and we knew that our marketing strategy would have to be flexible to fit changing circumstances. They told us that the U.S. Air Force Electronic Systems Division at Hanscom Field would be responsible for buying the ground stations and the commercial receivers. The Avionics Systems Division would only be responsible for specifying and buying the special military receivers. Even that activity, they told us, might be transferred to their counterparts at Hanscom Field.

We had worked with the people at Wright–Patterson Air Force Base for many years on several projects, but the Electronics Systems Division at Hanscom Field was new territory for us. We would have to visit the division soon to hear its side of the story.

We started to plan our campaign.

Intelligence Gathering

Intelligence gathering is a prerequisite to understanding every aspect of a competition. It means gathering information about the customer's requirements, the schedule, and competitive factors, including politics, budgets, and much more. Without accurate precompetition information, a company cannot develop and execute a winning strategy.

Throughout our campaign the marketing people working for C.M.C. Electronics guided us. Those people had U.S. government procurement experience, and because they understood the U.S. budget process, they also followed events on Capitol Hill. Before we visited Hanscom Field, we asked them to obtain more information about the program.

While they were collecting that information, we visited with Canadian government personnel. When Canada and the United States signed the Canada–United States Defense Production Sharing Arrangement in 1959, it granted defense contractors the right to bid for defense contracts in both countries on an equal basis. Despite the importance of the arrangement, we had learned that many people within the U.S. defense establishments did not understand its terms. For that and other reasons, we tried to keep our government personnel continuously informed of our progress throughout a competition for a U.S. defense contract. We did not expect them to try to influence the competition. We wanted them to be informed in case we needed their help to explain the bilateral arrangement. In that and other matters we did not leave anything to chance.

On Aug. 30, Tony Murfin, Carl Hanes, Sohel Fares, and I visited the U.S. Air Force Electronic Systems Division at Hanscom Field. It was the first of many trips we would make over the years to the Boston region.

The Air Force personnel surprised us when they said that they expected to release the Request For a Proposal in the spring of 1985, which was a full year earlier than the schedule we had heard of during our visit to Wright–Patterson Air Force Base and less than a year away. They expected to put a draft version of the receiver specification in their library in October, and they encouraged us to review it and provide them with our comments. That would be our first opportunity to help the Air Force produce a specification that met all its needs. It was also our opportunity to help make sure that the specification did not contain any unrealistic technical requirements. The process of helping customers write their specifications is an essential part of any competition strategy.

We discussed with the Air Force personnel the changes they would want the supplier to make to its airline Microwave Landing System receivers. They wanted a special military electrical interface circuit added. More important to us, they wanted a receiver that could provide all the features that were possible with this new landing system. We had already decided to design such a receiver. Our decision was risky because it meant that our equipment would be more expensive than the first-generation receivers now being built by Bendix and Sperry, though we expected those companies to design more elaborate equipment once the airlines and the military more clearly identified their needs. For the moment, we had a technical advantage for those markets.

The Air Force personnel asked us if we would provide them with an estimate of the costs for the modifications they wanted. They were trying to prepare their budget and wanted accurate information. They would pay for the changes but not for the design of the commercial receiver.

During the next three months we responded to the Air Force's questions and began to prepare ourselves for the upcoming competition.

We returned to Wright–Patterson Air Force Base because the personnel there were preparing the competition for one of the other receivers that the Air Force intended to buy. It was a special receiver that would be able to withstand the rigors of a fighter aircraft environment, as well as to operate from both the Instrument Landing System and Microwave Landing System ground stations.

We knew this military receiver program would be a long one. The program looked as if it would have several phases, during which competing equipment would be dropped. The Air Force expected to pay for most of the design work, but it could be eight years before it made a final selection.

The large and powerful Collins Avionics Division of Rockwell International approached Canadian Marconi to find out if we would be interested in a teaming arrangement. We had the all-important Microwave Landing System technology, and it had been building Instrument Landing System receivers for many years. It was also a large and respected U.S. corporation with many DOD contracts.

We spent many months making the trek between Montreal and Cedar Rapids, Iowa, where Rockwell-Collins is located. Finally, we reached an agreement. The result was a very successful collaborative effort between a giant in the U.S. avionics industry and a medium-sized Canadian company. This arrangement between two unlikely partners was possible because each had an essential role. Partnerships of this type work only when each participant can make a meaningful contribution.

By December 1984, we had most of the large Microwave Landing Systems programs covered. During the previous nine months we had begun to understand the U.S. triservice procurement program, that would lead the U.S. military to buy three different types of receivers.

With our commercial airline receiver and a joint development program with Rockwell International, we had two of the three covered. Earlier, we had decided not to try to offer the very special third receiver that the U.S. Navy wanted. It was already paying a company to design prototype equipment, and we did not think that we could influence and win that program.

The airlines were still proceeding with caution. However, we did help to persuade the Airlines Electronic Engineering Committee to begin work on an airline standard Microwave Landing System receiver specification. That work, which would eventually help us with the Air Force, would begin in March, but it was still only December.

The U.S. Air Force schedule began to slip, and the ground rules for the competition changed. The original plans, which called for the Air Force to buy equipment for the three services, did not hold. The Air Force had the most immediate need, and so they decided to proceed with competitions for two types of receivers: the modified commercial receiver for its transport airplanes and the special military receiver for the fighter aircraft. Those competitions, the Italian project, and others occurred at the same time.

Now I am going to break a rule I set for myself when I started this book—I am going to start using acronyms for systems, but only two. The first acronym, MLS, stands for the Microwave Landing System. I use it when I am referring to the new aircraft landing system that will be used by the world's civilian and military aviators. The second, CMLSA, stands for Commercial Microwave Landing System Avionics, and refers to the U.S. Air Force program to buy a modified airline-type MLS receiver.

In December and January, I wrote two memorandums describing the U.S. Air Force MLS program and our initial thoughts on how we would approach the competition. Tony Murfin and I would write many memorandums on that subject during the next three years.

We distributed those memorandums to all the people whose help we would need. Our purpose was to keep our colleagues and senior management informed about the program and our strategies as they unfolded. We knew that informed colleagues provide the most valuable help. We also knew that we would need the help of many experts, and so those memorandums went to people in our engineering, marketing, pricing, estimating, product support, publications, product assurance, and other departments. By involving those people from the beginning, we made them part of the team. We kept our corporate management informed every step of the way. All were essential to our final success, and their early involvement was necessary.

The Air Force decided that the ground station manufacturers should buy the receivers on its behalf. That meant that we would have to sell our receivers to the firms offering ground stations, and not directly to the Air Force. That was not a situation that we liked. We expected three receiver manufacturers and five companies to bid for the ground stations. We felt that we had a better chance in a head-to-head competition with the other receiver manufacturers with the Air Force as the evaluator.

The Air Force Electronic Systems Division hoped to release the Request for a Proposal during the summer or early autumn of 1985. We did not have much time for a major campaign that would require teaming agreements with one or more prime contractors.

A Strategic Campaign

In January 1985, about eight months before we expected the Request for a Proposal, I issued our first strategic plan. It included scheduling visits to the Air Force at Hanscom Field, visits to the companies we expected to bid for the ground station contract, visits to the Air Force user commands and several Canadian government officials, an advertising campaign, and more.

To succeed, we needed someone who could provide local liaison with the Air Force personnel. We tasked Tom Rieker, the marketing manager of C.M.C. Electronics, Inc., and its Washington, DC, representative, Tom Arnold, to look for such a person.

By February they had identified three people. Ervin Spinner and I went to Boston to interview the candidates. The third person we interviewed had an office on the second floor of the terminal building of the small municipal airport that was just outside the perimeter of Hanscom Field. When Ervin and I entered the office and met Dan Sullivan, we saw, as Ervin later described it, a movie set from Hollywood central casting for a private investigator's office.

Dan, a large jovial Boston Irishman, sat behind an old desk with a telephone on it and nothing else. Opposite the desk were two very worn chairs, and in the corner stood a solitary four-drawer metal filing cabinet. The room needed a coat of paint.

After 30 minutes with Dan, we knew that we had found our man. He had integrity, knowledge, and experience. Dan started by telling us that the information that he would obtain for us would be the type that was available to anyone, that is, to anyone who took the time to talk to our prospective customer. He was well known by most people in the Electronic Systems Division, but he did not claim to have any special privileges. What he offered was timely information, and lots of it, so that we could develop and execute a winning strategy. He insisted that we agree to visit Hanscom Field often. He knew about the need to maintain contact with prospective customers, to understand their requirements, and to help them prepare for the final competition.

During January and February we visited the Air Force and the five ground station bidders. We tried, unsuccessfully, to persuade the Air Force to separate the MLS receiver competition from that for the ground stations. Several of the ground station manufacturers told us they would want an exclusive arrangement. That meant we could only team with one of the five manufacturers, and our business would rely on that one winning the competition.

We prepared a briefing presentation about the program to educate all the people whose help we might need. After several meetings with people in the Air Force MLS project office, we began to better understand their special requirements, and we offered several suggestions to improve their specification. Then we began to change our own MLS receiver development program so that our product would meet their needs.

As the design progressed, its cost also escalated. Our receiver was complex because we had decided that it should be capable of providing all the features available from this new landing system. Although our design met the requirements of the Air Force and, we hoped, the future needs of the airlines, we began to worry that our receiver's cost might exceed its market value. Customers put a value on certain products and our receiver was starting to exceed the value of an aircraft landing system. We had to close the gap.

Tony led our team in a series of cost busting exercises that analyzed every aspect of the design and its manufacture. Those exercises became a regular occurrence for several years, as we strove continuously to reduce the cost of our receiver. Sometimes those design reviews led us to redesign parts of the equipment, but we had to accept those investments to ensure that we had a competitive product. Of all our avionics products, the MLS receiver proved to be the greatest cost containment challenge because we constantly had to face the reality that the product had a value to the customer, despite its superior performance, compared with earlier landing systems.

In March 1985, we held our first all-hands CMLSA planning and strategy session. By then we had visited all the likely ground station bidders and the Air Force several times. To be sure we gained the most from those visits, we took with us a list of carefully thought-out questions. We did not want to leave anything to chance.

By reviewing the draft technical specifications with the Air Force, we knew enough to begin writing our proposal. Once a Request for a Proposal is released, there is usually not enough time to put together a well-written proposal. Successful companies know to start that process early and then use the normal 30-day response time to refine their proposal based on the final bid instructions.

We decided to continue to try to convince the Air Force to separate the ground station competition from that for the receiver. Also, we had written to the Air Force MLS Project Office at Hanscom Field advising it that the Airlines Electronic Engineering Committee had started writing an MLS receiver specification. We suggested that the Air Force adopt that specification to ensure that the receivers it bought were compatible with those to be used by the airlines. In our opinion, the Air Force could save itself a lot of work by basing its own specification on the airlines' specification.

Because we were already designing an airline-type receiver, that strategy also would mean that we would not have to design equipment just for the Air Force. That was our best chance to be competitive. Despite all their other work, Tony Murfin and Carl Hanes began to participate in the monthly committee meetings to write what would become known as the ARINC-727 MLS Receiver Characteristic. We knew the process well from our Omega specification writing days.

At our all-hands strategy session, we drew up a comprehensive marketing action plan, but more importantly we went through the process of making the critical bid or no-bid decision. There was no point investing in an expensive campaign if we did not think that we could win. Because we had accumulated enough market intelligence information, we could make a rational analysis of our situation, our chances of winning, and what we had to do to win—all key elements for a bid or no-bid decision.

We critically examined our product as well as our engineering team. Did we have our best people on the project? The answer was yes. We compared ourselves with what we knew about the competition, and we rated the ground station bidders, the companies with whom we would have to team. We asked ourselves questions about our readiness to compete, the competitiveness of our costs, and our pricing strategy. When developing pricing tactics, companies have to decide if the competition justifies a long-term investment strategy. Then they must consider the overall state and financial health of their business. Other factors that we considered related to how well we understood the Air Force requirements, our situation as a foreign supplier, and our relationship with the Air Force MLS team.

After we rated the answers to those and many other questions, we were ready to decide if we could win. We concluded that we could, but we had much work ahead of us to improve our position. The major work was to stay close to our prospective customers and to help them whenever possible.

During March, we visited with buyers and C-130 users at the Air Force Warner Robbins Air Logistics Center. We briefed them on our MLS receiver program and our suggestion that the Air Force adopt the airlines' MLS receiver specification as well as buy the receivers through a separate competition. They took our suggestions under advisement. The program managers at the Electronic Systems Division were the only ones who could change the strategy.

By April we were well prepared for the upcoming competition, which we still expected in the summer or autumn. All we had to do was maintain our campaign by keeping in contact with the Air Force and the prime bidders, preparing our bid, and helping to write the airlines' MLS receiver specification. Then the U.S. budget season began in earnest, and the program started to fall apart.

Staying Close to the Program

Any company intending to bid on a U.S. government contract must understand and follow its budget process. It is a long and complex series of events that start nearly two years before the beginning of the U.S. fiscal year for which the funds apply. The fiscal year starts on Oct. 1 each year. The United States government sets budgets annually even for projects that will span several years. Understanding and following the process helps companies to plan and adjust their strategies, which is a key part of a winning campaign.

For our purpose, the U.S. budget can be split into two parts. The first and largest portion is those funds that the government calls entitlements. As the name implies, those are payments to which people and institutions are entitled. They cover social security payments, interest on loans, and similar nondiscretionary payments. The other part of the budget is those funds for agencies or projects that are not essential or are discretionary. The discretionary part of the budget, only about 35 percent of the total, has been the one to which the President and

Teaming a Product and a Global Market

Congress give the closest scrutiny, although that is changing to include the entitlements. The DOD budget falls into the discretionary funds.

The process starts with the various agencies preparing their budgets for the funds they will want two years ahead. About one year before the start of the fiscal year, the agencies submit their budget requests to the President's Office of Management and Budget. The President must submit his budget to Congress in January. Congress must approve the budget, and it usually makes many changes to the one proposed by the President. By April the budget season is well under way.

Programs have to be authorized and the spending of their funds approved, which is called the authorization and appropriation process. Congress has several authorization and appropriation committees that have the task of analyzing, changing, and then approving specific budget requests. The first task for budget watchers is to understand which committees are dealing with the part of the budget in which they are interested. Both the House of Representatives and the Senate have budget committees that sit separately. Eventually, both the House and Senate have to agree on the various items in the budget, which usually results in some form of compromise.

The committees hold hearings as they debate the various budgets, during which they ask the agencies to justify their requests. These debates continue until the budget is agreed on by all in the Congress and the President, usually for six months, between April and September.

The Air Force included $7.8 million for its MLS program as part of an overall air traffic control program, which they estimated would need $29.5 million between October 1985 and September 1986. In May the House of Representatives Armed Services Committee issued its report. It read in part:

> "The committee recommends a reduction [of] $7.803 million from the Air Force request of $29.517 million for the Traffic Control/Approach/Landing system.... The committee has historically opposed Department of Defense requests to fund expensive research and development programs to develop microwave landing systems.... The Air Force now estimates that development of a TMLS [Tactical MLS] will cost over $120 million....The committee believes that a great deal of money can be saved if the Air Force evolves a procurement approach that takes advantage of existing Federal Aviation Agency [Administration] landing systems [MLS] to meet the majority of the military requirements.....Accordingly, the committee recommends termination of the Air Force TMLS program as presently structured."

By the end of July, the conferees of the House and Senate authorization committees agreed with the earlier recommendation to disapprove the Air Force MLS program "as presently structured." The Air Force tried to counter that recommendation, still hoping that it could overturn the decision through the Appropriation Committee process.

Congress' problem was that the Air Force wanted to fund a company to design a special portable MLS ground station that weighed 500 pounds. The standard commercial ground stations weighed twice as much. The debate centered on a $7.8 million request within a multibillion dollar defense budget. It shows how closely Congress reviews all discretionary budget items.

While those events unfolded, we continued with our campaign. Early in the year Tony and his team had flown our receiver in the company's test airplane. In June the Canadian Armed Forces flew the receiver in one of its C-130 airplanes. We were anxious to fly in a C-130 to show the U.S. Air Force, as well as the Canadian forces, that our equipment would work in a military transport airplane. It was part of our strategy to build our credibility.

Tony led more cost busting exercises to find ways to reduce the cost of our receiver. The team also had to deal with many technical problems that surfaced as our design progressed. For years we had understood the advantages of involving our manufacturing and product quality experts during the design process. It was a tough challenge for Tony and his team, especially because we were spending much of our time positioning ourselves to win the Air Force program.

We continued to visit the personnel at the Air Force Electronic Systems Division to review and to comment on various drafts of their MLS specifications, and to tell them about our progress with our MLS receiver design.

Carl Hanes thought up the catchy name Microlander for our CMA-2000 MLS receiver. Later we produced an advertisement that featured a photograph of a C-130 airplane landing on a desert airstrip. Over the photograph we superimposed the letters MLS. The caption said, "We have the right equipment for Modified Commercial Microwave Landing System Avionics." We deliberately targeted that advertisement to the Air Force CMLSA program, and it highlighted the product's advanced technology and the features we knew the Air Force wanted. Whenever we visited the Air Force, we took with us a prototype receiver. We used every opportunity to show the Air Force that we had hardware and that we were designing an MLS receiver to its requirements.

To cover every aspect of the Air Force MLS program, we visited Scott Air Force Base, which is the headquarters of the Military Airlift Command, the operators of the C-130 aircraft and the principal Air Force user of the MLS.

By September the Air Force, as directed by Congress, was rushing to restructure its MLS procurement program. The Air Force MLS Program Office sent us a letter requesting that we provide it with detailed cost information to help it, once again, refine its budget estimates. In the request, it asked that we provide actual cost and other information about MLS receivers or similar equipment that we had in production.

Because of the competitiveness of the situation, we had difficulty deciding how to respond, but we knew we had to respond. Helping a prospective customer is important in any competition. Because our receiver was still in the design stage, we did not have actual cost data. However, our small Omega Navigation System was a comparable product, and so we provided information about its cost, reliability, and design.

The program continued this way for another year. We experienced times when we expected the Request for a Proposal within four months, and our adrenaline would flow. Then the Air Force would announce another delay. The specifications changed, and the program was restructured. Always, we stayed close to our prospective customer and worked to refine our proposal, our receiver design, its cost, and its pricing.

Finally, the Air Force decided to issue separate Request for Proposals for the receivers and the ground stations. It also decided to request receivers that conformed to the airlines ARINC-727 MLS Receiver Characteristic, a specification we were still helping the airlines to write.

Our optimism grew when the Air Force put in place our two primary suggestions, and we remained convinced that it was the best strategy for the Air Force. With that approach it was assured of a head-to-head competition among at least those manufacturers that were designing equipment for the airline market.

Tony Murfin suggested that the airlines include some of the Air Force MLS receiver requirements in their specification. They agreed, and that change allowed the Air Force to take full advantage of the airline specification. Also, it allowed the receiver manufacturers to design one product for both markets.

By the summer of 1986, two years after we started our campaign, the Air Force was confident that it would release its Request for a Proposal in the autumn. Now we focused most of our efforts on one goal—to win.

The Proposal Stage

During the summer of 1986, as we were about to embark on what we hoped would be the last stage in the Air Force CMLSA program—a program that had already had several false starts—we received some welcome news. Boeing, in consultation with a special branch of the Air Force, told us it had selected our MLS receiver for two new Air Force One airplanes. It was a major coup for a Canadian company to secure that prestigious contract, and winning it had been part of our overall CMLSA strategy.

The U.S. government had contracted with Boeing to provide two B-747 airplanes for Air Force One to replace the B-707s that the President was using. Boeing only required six MLS receivers, but we felt that the prestige was more valuable than the number of systems. Boeing did not need the receivers until the following year, but the order meant that we had to accelerate parts of our design work just when we were about to finish our CMLSA proposal. Once again, we asked our team, and Jean-Pierre Lepage, Tony's MLS project engineer, in particular, to work more hours than we could reasonably expect.

As with the Omega team, our MLS team had optimism, energy, and dedication. We had worked so closely with the Air Force that we were familiar with its technical requirements, and we did not expect any surprises when it released the Request for a Proposal. It was now time to refine our pricing strategy.

Someone once said, "Pricing is a tactic, not just arithmetic. You must price what the market will bear." Dan Sullivan, our man at Hanscom Field, recommended that we hire a company to carry out a should-cost analysis.

Some companies hire firms that specialize in writing proposals. Those consultants know how to respond to requirements in a Request for a Proposal to help ensure the bidder covers all points. I do not favor such outside help. If a company has not followed events leading up to a competition in enough detail that it needs help in writing its proposal, it should not bid the job. Similarly, a company's marketing department should not be given the task of writing an entire proposal. Experts in the various product disciplines should write the proposal. Once written, the proposal should be reviewed against the Request for the Proposal by an independent group of people. We usually refer to that in-house proposal review team as the Red Team. We followed this process for most major bids.

The pricing strategy, though, is a different matter. We felt that Dan's suggestion had merit. What we needed to know was what the Air Force might expect to pay for a modified commercial MLS receiver. Obviously, we were not privy to Air Force confidential information. Even an attempt to secure that type of information is illegal.

To better understand what the Air Force considered a fair price for the MLS receiver, we hired a company to carry out a should-cost analysis. It performed the analysis with information that we provided about our equipment design and labor rates and by studying the Air Force CMLSA draft specifications and previous similar Air Force contracts. Other than some of the information that we provided about our own equipment, the rest was publicly available.

That company produced several versions of its should-cost analysis, including one after the Air Force released its formal Request for a Proposal in December. The analysis provided us with guidance about what might be reasonable to bid for the special design work and the production receivers. It was a big help to us as we refined our bid, but it did not tell us what to bid to win.

The Request for a Proposal arrived on Dec. 17, 1986, two and one-half years after we started our campaign. When we received the Request for a Proposal, so late in December, we knew immediately that we would have to cancel Christmas.

We were about to bid on an approximate $20 million program that included a development phase with options for production contracts. Our bid had to be firm and fixed, which meant we had to have our costs right or go bankrupt—there would be no more money from the Air Force. We had to consider what might happen to the Canadian–U.S. exchange rate and other costs during the several years of the entire program. We frequently encountered this situation, but it always gave us cause for concern.

The request, as we expected, did not contain any surprises—we had done our homework. We had 45 days to submit the proposal. We had already written a draft technical proposal and much of the management proposal. Those parts would need updating, especially the technical proposal, because at the last moment the Air Force had inserted a limit of 80 pages for the document. We had much of our pricing strategy worked out, but we had the commercial proposal and several other parts still to write. Also, we had kept our senior management informed at every stage of our campaign, and we had their provisional approval for our approach and pricing strategy.

Two days after we received the formal request, Tony and I were ready for our bid kickoff meeting. We had not slept for two days as we read the Request for a Proposal, dissected it, and assigned writing and other tasks to all whose help we needed. Then we met with all our colleagues to explain the program once more, outline the Request for a Proposal, and distribute proposal assignments.

For the next several weeks we worked 16-hour days writing and refining our proposal. Though we had done a lot of preparatory work, we did not want to leave anything to chance.

I recall one Friday night when Stephen Joffrey, our pricing manager, and I filled the eight-foot-long blackboard in my office and many sheets of paper with new pricing strategies. We left the office at about 3 a.m., confident that we had produced a winning profitable bid.

As the proposal writing continued, we participated in the usual round of questions and answers with the Air Force. It is normal for a buyer to accept questions to clarify aspects of its Request for a Proposal. With the U.S. government, that process is structured and open. All bidders receive copies of the questions submitted by their competitors and the answers. Attentive bidders try to estimate how prepared their competitors are by the questions they ask. As I remember it, we did not ask any questions.

One result of the questions and answers was that the Air Force extended the date for the delivery of our proposals. The big day finally came, and, as usual, we hand-carried our proposal to our prospective customer. Left behind was an exhausted team. The anxious wait started.

In the months ahead there were more questions and answers. This time the questions came from the Air Force evaluation team.

During the summer we received word from Dan Sullivan that the Air Force would most likely require a Best and Final Offer—the dreaded BAFO. Such an event was common in major bids of this type.

Once again we faced the task of analyzing our costs and our short- and long-term business strategies. I decided to go ahead with the annual family camping vacation while we waited for the Best and Final Offer request. Some vacations are just not meant to be. I spent more time in telephone booths than with my family as we traveled through southern Nova Scotia and then the rustic Mount

Desert Island in Maine. I continued to discuss strategies with Tony Murfin and Stephen Joffrey. I called so often from phone booths that Stephen started to call me Clark Kent. We cut our vacation short. I was too anxious about the CMLSA program.

The Best and Final Offer request finally arrived, we responded, and then there was more anxious waiting.

One beautiful day in October I received a call from our Boston Irishman, Dan Sullivan. Tony Murfin, who had the office next door to mine, rushed in as soon as he heard me say, "Hi Dan." Dan replied with two words: "We won."

19

They Did Not Become Complacent

The pattern of growth has been one of cautious aggressiveness. When a product type or technology is mastered and no longer entails major unknown risks, then new markets are explored for this technology. Conversely, when a market segment and its business practices have been mastered, then an attempt to venture into a new technology is made. Thus, by alternating advances in market and technology, sensible growth is achieved without exposing the company to potentially damaging risks.
—Ervin Spinner
Avionics Division Business Plan, 1983

For our Commercial Avionics Group to grow and to prosper, we needed new products. Because we had mastered the Omega System, we were confident about tackling any market, even with Omega products that used new technologies. We proceeded cautiously, though, with our other products.

The Fuel Management System presented us with a particular and new challenge, because it was unique to a particular type of airplane. Also, it was our first product that included the performance characteristics of an airplane.

The new airline People Express showed considerable interest in that product for its Boeing B-727 and B-737 airplanes, and so, Tony Murfin and his team concentrated on making the Fuel Management System compatible with the airline's particular airplanes. Because we did not yet have a contract with the airline, the work was at our own risk.

Much of that work, with its uncertain future, came when the other half of Tony's team was working on our MLS receiver, itself a full-time job. Because People Express was also interested in buying a large quantity of our Alpha-Omega

Navigation Systems, we had high expectations for People Express soon becoming a major customer.

We did not lack ideas for new products; our challenge was to select those that fitted a logical business strategy and whose design we could afford.

When he was administrator for the FAA, Lynn Helms said, "An organization must have a written plan for the future, only then can it know where it is going and get there."

I have always been a strong proponent of written plans, and it is preferable if they are written as a team effort. Then everyone has a feeling of ownership and commitment to the plan. Sometimes such participation is not possible, and then it is the manager's job to collect suggestions and write them down. The act of writing down a plan is also one way to think more carefully about its content. A written plan has a better chance of working than an unwritten one.

Already, I had a reputation as a prolific writer of business and marketing plans, and so my colleagues were not surprised when late in 1983 I decided it was time to write a business plan for our new Commercial Avionics Group. I issued the plan early in January 1983, and so it was another self-imposed Christmas homework assignment. I wanted to organize my own thoughts about our future, which I could do only by writing them down, and I wanted to share my thoughts with everyone within the group.

In the plan I wrote about products that we had, products we were committed to designing, and products we wanted to design. The plan also covered those military products being designed elsewhere in the division that we could modify for the civilian market.

What we had was a broad range of Omega Navigation Systems, with an MLS receiver in the design stage. Elsewhere in the division, engineers were designing a military navigation system that would use the U.S. Air Force Global Positioning System satellites. The Global Positioning System technology was one that we knew we should covert into a commercial avionics product.

We had a fuel management system, parts of which we felt we could use with our navigation technologies to give us a product that combined both navigation and flight management features. It would be a flight management system similar to but smaller than many on the market. We decided to call that system the CMA-900.

Several other ideas we had would allow us to offer a range of navigation and flight management systems that would cover the entire flight profile from takeoff to touchdown. We developed that theme in several advertisements.

Now we had a logical growth path for navigation products. Most were interrelated, which had been part of our strategy to reduce our investment while taking full advantage of the work that we could afford.

Another part of avionics on which we started to focus was aircraft monitoring and display systems. The division already had a very successful line of engine instruments, and we were beginning to sell those into business aviation.

Tony's team had inherited an aircraft equipment monitoring system, and Sol's group in Kanata was working on an engine health monitor. That was another product idea in which the airlines were becoming increasingly interested.

We had the beginnings, by combining products, of an aircraft equipment and engine health monitor within one box.

Our long-range plan charted a course to bring us to a point where we could offer a range of related navigation and flight management system products. It also showed the way to a group of products that would help a pilot know if any part of the airplane was not working correctly. It was a bold plan for a company that only 10 years earlier had started the transition into the civilian market. With some changes, we put that plan into place.

However, I was not quite finished with our planning. The airlines were becoming increasingly interested in communicating with the ground through a digital not just a voice link. Initially, they wanted to automatically monitor the performance of aircraft equipment and to send that information to the ground crews. Then technicians on the ground could analyze the information to find out if anything had broken and could be waiting with a spare part when the airplane landed. Previously, they had had to wait until an airplane landed to find out what might need replacing. With an air-to-ground digital data link, the airlines also could begin to plan to provide an in-flight telephone service for their passengers.

Putting a system in place over land was not that difficult; what interested me was the challenge of flights over water Here the only form of communication was a long-range high-frequency communication system.

My attention to that problem was not due to any special talents. The FAA and the International Civil Aviation Authority also wanted a digital data link with airplanes on routes over water. They wanted to make air travel safer, and the job of the ground controllers easier, by having an airplane's position reported more frequently and automatically. The usual procedure was for pilots to communicate their position periodically over the high frequency voice communications system. A digital data link working with the airplane's navigation system could provide that information more frequently and automatically.

Don Mactaggart, our first Omega product champion, was beginning to tire of his latest championship, our telex exchange business, and so he started to drift in and out of my office. He was looking for a new and radical challenge. Why not? "Don, I've got this idea."

Within weeks, in early 1985, Don had put together an engineering concept to use the high-frequency communication network for a digital data link. It would need repeater ground stations, a gateway exchange, and perhaps even some land lines. It was radical, but Don immediately started talking to Air Canada, other airlines, and some of his contacts in the communications industry. I tried to keep up with him. Meanwhile, the airlines and others in the industry started to look at what it would take to set up an airline satellite communications service.

Canadian Marconi had been involved in a satellite communication research project with Transport Canada, our aviation regulatory agency, during the early 1970s. For that project Dave Allcock, who was now working on our MLS receiver design, had led a team in the design of an airplane satellite communications antenna, and so we had some experience with that type of product.

For many months Don and I immersed ourselves in trying to convince our corporate management and the airline industry to support our concept for an aircraft high-frequency digital communications system. Don, the radical product champion with many unorthodox habits, was back—it was fun.

Our concept was a little too radical for most, but our work heightened interest within Canadian Marconi for a digital data link product. By 1987 the airlines were beginning to plan to use a satellite communications system. Because of our experience and our newfound knowledge, I promoted a project to design a satellite communications system, which would be a logical addition to our growing list of civilian avionics. We started by designing an airborne satellite antenna, a very complex and technically challenging product. It was also the last product that I helped to launch at Canadian Marconi.

By 1988 the company was well on its way to successfully completing its transition into the civilian market.

Since I have lived in Washington, DC, the only pieces of mail that I read immediately are two Canadian Marconi newsletters: *CIRCUIT*, the company's in-house newsletter, and *CONTACT*, a newsletter for general distribution. Yanka Dvornik, their editor and the company's manager for corporate communications, succeeds in cramming more interesting articles into those two newsletters than many editors of national newspapers manage to put into their publications. Yanka's talents, cheerful infectious enthusiasm, and editorial skills make her a corporate manager's best friend. Communication is essential for any successful endeavor, and many companies would benefit from following Yanka's model for corporate communication.

The October 1990 edition of *CONTACT* contained several articles that illustrated the success of the commercial avionics product design strategy that we had followed. On April 10, 1992, the spring edition of *CONTACT* arrived with four articles about recent orders for the company's commercial avionics. Fifteen minutes later I knew that the company's transition into the civilian market was complete, and immediately I knew how I wanted to finish this book: with quotes of passages from a number of articles that appeared in several issues of the company's two newsletters.

"Canadian Marconi Company's Commercial Avionics: Expansion Lies in Complementary Products," CONTACT, Fall/Winter 1990

After being a one-product group for many years, Commercial Avionics is building upon its solid technical base of Omega products to include such technologies as GPS [satellite navigation equipment using the Global Positioning System], MLS [aircraft landing equipment using the MLS], Satcom [satellite communications equipment], and LCD Display Units [cockpit control and display units that use liquid crystal display technology] which are featured in the accompanying articles.

According to Henry Schlachta, Manager - Commercial Avionics, "The approach being taken with such diversification is multi-product marketing. That is, an integrated package of our avionics products which complement each other from takeoff to touchdown. . . ."

Despite the differences between the military and civil markets, both share a high demand for reputation, product support, price competitiveness and quality technology. The commercial market, however, is marked by a much shorter lead time of 30 days or less for delivery of the product.

This reality entails maintaining a strong, ongoing relationship with the customer so as not only to anticipate his needs but, more importantly, to predict when those needs will come into play so as to be prepared with off-the-shelf equipment.

"The GPS agreement with Universal is a prime example of how such readiness wins contracts," adds David Bailey, Commercial Marketing Group Manager - Avionics. . . .

[Universal Navigation Incorporated is a company that has been very successful in selling its flight management system to business aviation companies. First we sold our Omega sensor to Universal for use with its own product, and then the company received a large contract to supply Universal with Omega and Global Positioning System sensors combined in one small unit. Also, this was another example of our willingness to work with companies that had a competing product. We had a flight management system, our CMA-900, that competed with Universal's, but the sensor business was one we could not ignore.]

"I would also highlight product support as being essential to commercial customers, the logic being that your product is only as strong as the resources backing it up." . . . Future success in the commercial arena is contingent upon foreseeing a customer's particular needs and responding to them expediently with diverse and reliable products and services. The strategy to repeat this process through product line expansions is just one of the many ways in which CMC will keep growing both with its clients and in the marketplace.

"Omega Products: 'Bread and Butter' of Commercial Avionics," CONTACT, Fall/Winter 1990

Twenty years and four equipment generations after the first flight trials of the CMA-719 Omega Navigation System back in August 1970, Omega products continue to bolster Canadian Marconi Company's commercial success with total sales exceeding $150 million. "With orders to date for 4000 systems, the Omega product line is undoubtedly our 'bread and butter,'" says David Bailey, Commercial Marketing Group Manager - Avionics.

A major factor in CMC's success has been the continuing evolution of the Omega line to encompass a complete family of products that is suitable for a wide range of applications including airliners, corporate aircraft, helicopters and military transports.

The [CMC's] strongest growth in recent years has been in the corporate aviation and airline sectors. In the former, the CMA-764 continues to set the industry standard for long-range navigation sensors [the small Omega product that we started in 1985 using the most advanced component technologies]. Success in the airline sector is anchored by the position of the CMA-771 Omega Navigation System [our airline standard equipment that complies with the airlines ARINC-599 Characteristic], as a sole factory option on McDonnell Douglas MD-80 series aircraft.

The prosperity of the Omega System [product line] is assured well into the next century . . . by offering [in 1991] embedded Global Positioning System receivers in the CMA-764 and CMA-771. . . .

"First Flight," CONTACT, Fall/Winter 1990

Westland Helicopter's EH101 Heliliner promises to set new standards for commercial helicopters. Last April 24, this civil version of the EH101 performed its first successful flight aided by two new CMC [Commercial] Avionics products, . . . the CMA-2014 Multi-Purpose Control and Display Unit and the CMA-900 Flight Management Computer. . . .

"CMA-2000 airborne MLS: 'A Major Opportunity in the Commercial World'", CONTACT, Fall/Winter 1990:

In the last few months, Canadian Marconi's CMA-2000 airborne MLS [receiver] has been performing successfully in commercial and government test trials, bringing it one step closer to certification on McDonnell Douglas's new MD-11 aircraft. . . . According to Tony Murfin, CMC Product Manager - Flight Systems, the MD-11 program represents "a major opportunity in the commercial world" for the CMC airborne MLS.

The first successful CMA-2000 testing dedicated to the MD-11 program occurred at a Honeywell Inc. Validation Facility in Phoenix, Arizona. Honeywell is the prime supplier of most of the cockpit avionics for the MD-11.

"New-Generation GPS Selected by United Airlines," CIRCUIT, December 1992

The team of Canadian Marconi Company and Honeywell's Commercial Flight Systems Group has recently been selected by United Airlines to supply Global Navigation Satellite Sensor Units for United's Boeing B-777 aircraft. . . .

[The Canadian Marconi Company completes its transition into the commercial market.]

"Canadian Marconi Awarded World's First Major Airlines Production Contract," CONTACT, Spring 1992

Canadian Marconi Company has recently won the final phase of a U.S. Air Force contract worth $18.4 million, for the supply of airborne MLS receivers which enable aircraft to land safely in extremely bad weather and low visibility.

The contract calls for the provision of over 1,000 AN/ARN-152 receivers as part of the USAF's Commercial MLS Avionics program. This is the world's first major production contract awarded to an MLS manufacturer. . . .

The AN/ARN-152 is the result of over 10 years of design and development at CMC, and is the militarized version of the company's CMA-2000 [commercial] MLS receiver. . . .

Canadian Marconi has also recently delivered a combined MLS/ILS [Instrument Landing System] airborne receiver, in conjunction with the Collins Avionics and Communications Division of Rockwell International, as part of the USAF Military MLS Avionics program.

"CMC's New-Generation GPS Selected for Boeing 777," CONTACT, Spring 1992

The team of Canadian Marconi Company and Honeywell's Commercial Flight Systems Group has recently been selected by Boeing as one of two manufacturers whose Global Positioning System Sensor equipment will be a standard option on Boeing's new B-777 aircraft. . . . Honeywell is a world leader in the commercial avionics field. . . .CMC's more than 15 years experience in GPS [Global Positioning System] receiver design was a major factor in being selected first by Honeywell, followed by Boeing.

Under the CMC/Honeywell cooperative program CMC is fully responsible for the design and manufacture of a 12 channel GPS Sensor Unit to the ARINC-743A specification [another one that we helped to write]. . . . Several thousand units are expected to be produced by CMC and delivered to Honeywell over the next eight years. The result of a major CMC engineering effort, the new GPS product will achieve dramatic advances in performance, reliability and integration as compared to previous generations [of GPS products].

[That venture with Honeywell, like the military MLS joint venture between Canadian Marconi and Rockwell-Collins and others, is one more example of Canadian Marconi finding ways to work with major companies in foreign markets. The comparative sizes of companies need not be a deterrent to cooperation.]

"First Deliveries of New CMC NAV System with GPS," CONTACT, Spring 1992

Canadian Marconi Company has recently begun deliveries of its CMA-900-1 Navigation Management System for REGA Swissair Ambulance's A109K2 helicopter. . . .

The CMA-900-1 is a navigation management system which provides automatic enroute navigation based on data from a built-in commercial or C/A code GPS receiver and other external navigation sensor inputs. Navigation data is displayed on Canadian Marconi's multi-line, colour Control Display Unit, the CMA-2014 [the product whose design we started in 1985]

"Launch SatCom Order From McDonnell Douglas," CONTACT, Spring 1992

Canadian Marconi's new generation CMA-2102 SatCom high-gain antenna, now in advanced stages of development, has been selected by McDonnell Douglas as a standard option on the MD-11 aircraft. This is the launch order for the CMA-2102, following the sale of some 20 first-generation [aircraft satellite communication] antennas over the last two years.

"CMC SatCom Antenna Type Certified for B-767 Aircraft," CONTACT, Spring 1992

Following the successful completion of ground and flight tests, the CMA-2100 SatCom antenna has been granted FAA certification as part of a complex airborne terminal on a Boeing B-767-300 aircraft. . . . The CMA-2100 has been approved by the FAA for operation on B-747 and B-707 aircraft.

* * *

All the major new products that we started during our search for excellence had made it into the market.

Only I know how difficult was my decision to leave the Canadian Marconi Company. In the end, the compelling nature of the space business, the opportunity to represent Canada on the world's largest space program, and a new challenge were just enough to sway me.

The hardest part of my decision was knowing that I would be leaving the team. We had done so much together, and everyone had played a vital role. Total Quality Management, without the formal title, had worked for me. Canadian Marconi was, and remains, an important part of my life.

This book has been in many ways my way of recognizing the accomplishments of the entire team, a team that includes past and present Canadian Marconi employees, government officials, customers, competitors, and many others; they all participated in our search for excellence in Teaming a Product and a Global Market.

Epilogue

Was It Worth It?

There are many ways to measure worth or success. From the inside, working every day on the front line, the work often seems chaotic. Worse, as the markets and global economies continue to change, employees can become discouraged. They look for quick solutions and wonder about their future—a future over which they may feel they have little or no control—a future being dictated by corporate strategies. Some hardships are inevitable. The companies that survive the troughs of economic and market cycles, then grow as the opportunities once again materialize—as surely they will—are the ones that seize the initiative, respond to their customers, strive to balance their markets, and make a long-term commitment. Successful companies are ones whose corporate management listen and respond to those employees who are most intimately familiar with the markets. Today, we call this empowerment. As with so much that we did at Canadian Marconi, which today has a name tag, we empowered our employees because it seemed sensible.

While writing this book I tried to be objective about what we did right, what we did wrong, and what kept us going—in Teaming a Product and a Global Market. Being, now, a Canadian Marconi outsider has helped me to be objective. Those on the inside see and participate in the daily struggles to maintain excellence. However, the search for excellence should never stop, and the way forward requires hard work and sacrifice.

There is no doubt we succeeded in our goal to change from an exclusively military systems supplier to a balanced military and commercial systems company. While Canadian Marconi annual gross revenues have cycled between $300 million and $250 million, in 1995 commercial sales represented 46 percent of the company's revenues with 55 percent of those sales being aerospace products.

Those of us working to develop a global commercial market were continuously challenged, and for most of the time we enjoyed a strong corporate commitment to our goal. Without that long term commitment we would not have succeeded. As readers of this book will have seen, companies larger than ours did not stay the course and suffered as a consequence. From an employee's perspective it was an exhilarating period with much job satisfaction. However, did the investment justify the result?

Our shareholders and senior executives, especially those of our parent company, understandably focused, though not exclusively, on profitability. Annual, even monthly, profitability was carefully scrutinized. Though the early years were not profitable, our shareholders, public and corporate alike, allowed us to stay the course. However, and as a consequence of the need for an ongoing investment, along the way there were personnel casualties. I would like to think our corporate masters were thinking long term. If they had, they would now be able to look back and see that in the 24 years since 1972, the Omega product alone generated $200 million in sales. More than 5000 systems, spanning six generations of the product, were delivered mostly to commercial customers throughout the world. Profitability figures are privileged information; however, the return on investment on gross expenditures was better than 20 percent.

It was some six years before the product line became profitable. Our ability to continue with the investment resulted from several factors. The company had two or three profitable bread-and-butter product lines. These provided a predictable source of revenue and allowed the company to reinvest some 15 percent of its revenues in research and development. In the early years, research and development expenditures were augmented with government grants through what was then called the Defense Industry Productivity Program. Canadian Marconi benefited from a 50–50 sharing of the development costs for our first-generation Omega system. The government invested to a lesser extent in some of the later generations of the product line and this certainly contributed to our success. However, the company had to make substantial investments, not just in nonrecurring development costs, but also in substantial support start-up costs and a sustained worldwide marketing campaign.

Securing a prestigious and large launch customer—Pan American World Airways—was crucial to our success. This not only brought much needed revenues but renewed faith in our ability to succeed. Rather than strive only for profitability, we invested much time and energy into ensuring that Pan Am was a satisfied customer in every respect, and this paid dividends as we capitalized on the Pan Am contract to expand our customer base. Our strategy was to capture a large market share believing that profitability would follow. All our sales were profitable, but for most we contained our profitability to 10 percent to be able to invest in product support. Also, we responded to our customers' unique requirements. Within a particular generation of the product, there were many variants, all dictated by a customer's special need. Our willingness to modify a product or add a feature became one of the hallmarks that differentiated Canadian Marconi from the competition. This required a sustained investment.

I remain convinced that to successfully launch a new product or venture into a new market the venture needs a Product Champion and an organizational structure that allows the Product Champion to manage all aspects of the product's life cycle. For the era we were in, launching the Omega product line and venturing into the commercial market, we benefited from being organized around a product manager who had not just the responsibility and but also the authority.

One person cannot make a business successful. Nor should anyone try to claim that he or she alone is responsible, although the adage "success has many fathers and failure is an orphan" is sadly true. Success requires product champions, the entire team, and as we found out, the team includes many people outside the business unit, not just those on the inside. However, I credit two people for our success. Don Mactaggart showed vision in pursuing the potential of the Omega System even before the U.S. Navy was authorized to develop an operational ground network. He, with the support of, then, Avionics Division Vice President Kieth Glegg, continued with the product development even in the face of early corporate scepticism. I credit Peter Gasser as the person who launched the product line and contributed most to its success. He must be given credit for his technical aptitude during the initial development phase. He deserves credit for his technical vision when the time came to take the next step and develop a second-generation product. He provided technical and personnel development leadership when it was needed and he left the engineers alone to use their own skills when appropriate. From the all-important marketing perspective, he gave our potential customers confidence in his own abilities, those of the team, the company and our product. Through the years the Omega product line was blessed with a dedicated and professional team, and it would be an insult to more than 100 people to give all the credit to only one or two people—all deserve credit. Looking back I can see that because we put much faith in common sense and our people, we put the right people in the right place at the right time. It was purely coincidental that certain careers, my own included, were enhanced because we happened to be the right people, at the right place, at the right time. No personal agendas were pursued.

Was it worth it? The product line was a financial success when viewed over the long term, and we succeeded in securing a leadership position in a global commercial market. The success of our story, however, is probably best measured by looking at it from a people perspective. There is no doubt the company and its people benefited from Teaming a Product and a Global Market.

Index

Accuracy of systems, 35, 105, 121, 132–133, 159–160, 189–191
Advertising, 153, 146, 169, 177, 226, 232
Advisory Circulars, 97, 114
Aeritalia, 63, 129
AeroMexico, 132, 168
Aeroflot, 101, 166, 169, 203, 207
Aerolineas Argentinas, 53
Air Afrique, 194
Air Canada, 66, 91
Air Force, U.S., 26, 33, 41, 43–44, 77–80, 102, 128, 198, 215
    budget process, 225–227
    CMLSA program, 227–230
    Dynell contract, 166
    Electronic Systems Division, 218, 222, 224, 226
    jet transport aircraft, 45
    Logistics Command, 49
    Military Airlift Command, 48–49
    MLS receivers, 216–217, 237
    procurement process, 48–49, 216–230
    test flights, 67
    Warner Robbins Air Logistic Center, 214
Air Force One, 227
Air France, 66, 91, 101, 169
Air Guinea, 194
Air New Zealand, 141, 156, 168, 194
Air Pacific, 133, 138, 190, 194
Air traffic control, 214, 225
Airbus Industries, 173
Aircraft Electronics Association, 188, 216
Aircraft manufacturers, 122, 129, 184, 194
Aircraft on the ground (AOG), 189–190
Aircraft. *see* specific aircraft i.e. B-727 aircraft; C-130 transport aircraft
Airframe manufacturers, 56–57, 127, 156, 161, 167, 170, 175–177, 184, 208, 216
Airline industry, 175–177, 208
Airlines Electronic Engineering Committee, 90–91, 93, 98, 101, 104, 106, 107, 113, 116, 158, 164, 193, 223
Air-to-ground digital data link, 233
Allcock, Dave, 214–215, 234
Allende, Salvador, 52
Alphanumeric display, 193, 195, 196, 208, 209, 210

Alpha-Omega Navigation Systems, 13, 231–232
Altimeter, 81
American Airlines, 91, 101, 113, 126, 166
American Transair, 194
AN/ARN-99 system, 37, 46, 58, 72
AN/ARN-115 system, 42–43, 49
AN/ARN-152 system, 237
Antennas, 20, 63, 167, 186, 238
    airborne satellite, 234
    Argus aircraft, 41
    CMA-719, 24
    externally mounted, 74
    location of, 16, 124
    loop, 22, 23
    North America, 4
Appliance companies, 172
Arantes, Assis, 55–56, 63, 64, 121–122, 141, 159, 160, 161, 173, 190–191
Arctic region, 34, 177–178
Argentina, 15, 52, 53, 105
Argentinean Air Force, 53
Argus aircraft, 27, 41, 42, 43, 61
ARINC characteristics, 91
ARINC-580 Characteristic, 92, 93, 94, 97, 103, 104, 116
ARINC-599 Characteristic, 116, 158, 236
ARINC-727 MLS Characteristic, 223, 227
ARINC-743A specification, 238
Army, U.S., 5, 78, 215
    budget, 214
    contracts, 118
    helicopters, 81
    instrument orders, 145
Arnold, Tom, 222
Arrow-Omega, 13
Australia, 14, 15, 105, 138
Autopilot, 94, 116, 161–162
Avensa, 194
Aviaexport, 203
Aviation Week and Space Technology, 99–100

B-707 aircraft, 45, 47, 53, 56, 57, 61, 63, 65, 66, 100, 102, 103, 108, 114, 159, 190–191, 238

Index

B-727 aircraft, 44, 66, 96, 115, 131, 167, 169, 176, 186, 210, 202, 208–209, 231
B-737 aircraft, 44, 115, 131, 167, 178, 202, 208–209, 231
B-747 aircraft, 19, 44, 47, 100, 171, 173, 227, 238
B-757 aircraft, 176
B-767 aircraft, 176
B-767-300 aircraft, 238
B-777 aircraft, 237
Bailey, David, 32, 155, 198–199, 201, 215, 235, 236
Baillie, Robert, 61–62, 66, 72, 74, 124, 149, 155, 160
Banderante, 122, 129
Barker, Cliff, 95
Beech Aircraft, 121
Beechcraft King Air, 67
Bendix Corporation, 48, 62, 71, 78, 80, 103, 105, 123, 124
   as competitors, 109, 110, 111, 121, 132, 166
   ground stations, 215
   MLS receivers, 217, 219
   Omega design, 90, 94
Bercier, Gilles, 139, 190
Bertie, Tom, 52
Best and Final Offer (BAFO), 229
Bidding, 110–111, 134–135, 152, 158, 224
   analysis, 125
   approach to, 135–136
   BAFO, 229
   ground stations, 221–222
   pricing strategy, 228, 229
Black and Decker, 158–159
Boeing Commercial Airplane Company, 19, 27, 29, 57, 96, 115–116, 129, 131–132, 167, 173, 176, 186, 201, 202, 208, 227
Boileau, Gilbert, 20, 32, 72
The Borderless World, 172
Boss, John, 22
Boudreau, Eddy, 150
Bowyer, Don, 101
Braathens SAFE, 167–168
Brazil, 52, 62–64
Brazilian Air Force, 122, 188
British Airways, 101
British Civil Aviation Authority, 176
British Telecommunications, 118, 150
British West Indian Airways, 127–128, 156
Bruce, Jim, 175
Buker, Bill, 29, 199
Bureaucracies, 134
Business aviation, 67–69, 117, 121, 167, 172
Business jet operators, 216
The Businessman's Guide to Japan and Its Customs, 67
Butler Aviation, 69, 110, 181

C-130 transport aircraft, 41, 47, 48, 49, 61, 65, 108, 128–129, 131, 156, 166, 224, 226
C-141 transport aircraft, 47, 49, 58
Caden, Paul, 203, 204, 206
Canadair, 216
Canadair Challenger, 67, 202, 209
Canadarm, 8
Canada–United States Defense Production Sharing Arrangement, 88, 219
Canadian Armed Forces, 21, 27, 47, 186
   CMA-719 purchase, 97–98
   Maritime Proving and Evaluation Unit, 50
   Omega contract, 61–62
   Omega tests, 41–42, 43
   Request for Proposal, 61
Canadian Marconi Company
   Avionics Division, 5–6, 18–19, 30, 48, 72, 80, 83, 149–150, 178, 196–198, 218
   Avionics Product Support Group, 68–69
   business environment, 144
   Commercial Avionics Group, 148, 198–200, 215, 231–232
   Commercial Avionics Marketing Department, 151–152
   Commercial Flight Systems, 198–200, 215–216
   Commercial Navigation Systems, 199, 215–216
   communications, internal, 152
   Continuous Improvement Principle, 193
   early history, 3–8
   management, 149, 151
   Marketing Services Department, 153
   Omega Lab, 32, 34, 71, 98
   philosophy, 69, 72

reorganization of, 148–151
Training Department, 120–121
Canadian National Defense, 37, 41
Canadian Pacific Airlines, 194
Cape Cod, Massachusetts, 4
Caravelle aircraft, 27, 65
Carry-on systems, 178
Cathay Pacific, 133, 137, 168
Cathode-ray tubes, 81, 195–196
Centralized management, 7
Certification program, 176
Cessna Citation, 67
Chamberlain, Ormee, 41, 65, 120, 139
Champagne, Gilbert, 154
Che, John, 203–204
Chile, 52
Chilean Air Force, 52–53, 156, 162–163
China, 203–207
China Airlines, 137, 156, 168, 194
CIRCUIT, 234–238
Circuit boards, 25, 150
Circuit cards, 74, 94, 195
Civil Aviation Authority, United Kingdom, 93
Civilian markets, 19, 44, 47, 100, 123–124, 126–127, 133–134, 137, 146, 149, 171, 173, 181, 183–185, 198, 210, 216, 227, 235, 238
CMA-719 system, 24–28, 32, 41, 43, 58, 71–75, 93, 101–102, 155, 167, 186, 236
  development cost, 61
  flight tests, 34–35, 45–51, 57, 59, 62–67
  marketing of, 25–28, 44
  price, 210
  production, 34
  purchase by Canadian Armed Forces, 97–98
CMA-734 Arrow, 210
CMA-734 system, 71–75, 93–94, 98, 101, 106–107, 109, 114, 117, 119, 146, 155, 163, 176, 178, 186, 194, 196, 210
CMA-740 system, 94, 98, 101, 102, 105–106, 107, 109, 119, 158, 159, 163, 186, 187, 196, 210
  flight tests, 132
  manufacture of, 114
  Varig Brazilian Airlines, 121
CMA-759 system, 176
CMA-764 system, 211
CMA-771 Alpha–Omega System, 177, 187, 195, 196, 208–209, 210

CMA-771 system, 116, 117, 158, 163, 167, 176–177, 186, 202
CMA-802 system, 193, 194
CMA-860 system, 193
CMA-900 system, 232, 235, 236
CMA-900-1 system, 238
CMA-2000 system, 226, 236–237
CMA-2014 system, 236, 238
CMA-2102 system, 238
C.M.C. Electronics Incorporated, 110, 199, 216, 219
CMLSA. see Commercial Microwave Landing System Avionics
Coast Guard, U.S., 14, 22, 42–43, 59, 62, 194
Coastal zone management, 129
Coastguarder aircraft, 129, 131
Cockpits, 170, 186
Collins Avionics, 237
Collins Radio, 62
Collision detection and avoidance system, 209
Comet test aircraft, 21–23, 25, 47
Commercial aircraft, 19, 27, 42, 114, 115, 122–124
  see also specific aircraft
Commercial markets, 55–58, 60–61, 66–67, 91, 99–111, 113, 131, 143–145, 151–152, 155, 156, 162–163, 166–168, 194, 198, 199, 237
Commercial Microwave Landing System Avionics, 221
Communication signals, 12–13, 15, 20, 39–40, 79, 80, 107, 174, 186, 233
Communications broadcasting companies, 4
Communications Components Corporation, 40–41, 80, 104, 143, 146, 167
Communications satellite system, 209
Compass system, 162
Competition, 39–40, 48–49, 66, 71–73, 78, 80, 88, 92, 103–104, 109, 113, 121, 126, 133, 160, 166
  bidding, 110–111, 125, 135–136
  corporate business, 156
  information gathering, 218–230
  MLS, 220–221
  receivers, 219
  understanding of, 170
  see also names of specific competitors
Computer storage devices, 24–25

Consignment of parts, 141
CONTACT, 234–238
Continental Airlines, 45
Contracts, 168, 190, 194
   Air Force, U.S., 82
   Army, U.S., 118
   awards for, 138
   Canadian Armed Forces, 61–62
   competition for. *see* Competition
   defense, 219
   negotiations, 181, 203, 207
   Northrop/U.S. Navy, 20, 21, 26, 33, 41, 44–45, 88
   Omega Systems, 128–129
   Pan American World Airways, 111
   receivers, 217
   Varig Brazilian Airlines, 140–141
Contracts administration, 6, 200
Control and display units, 176–177, 195, 196, 210, 235, 236, 238
Corporate aviation, 67–69, 121, 143–144, 156, 236
Cost containment, 210–212
Court, Jim, 30, 31
Credibility, 64, 82, 96
Crouzet, 66, 80, 166–167
Cruzeiro, 194
Customer base. *see* Marketing; specific market
Customer's rights, 185–191

Dassault Falcon, 67
DC-3 aircraft, 25
DC-8 aircraft, 45, 47, 57, 66, 104–111, 132
DC-10 aircraft, 159, 173, 191, 194
De Havilland Buffalo transport aircraft, 61
De Havilland of Canada, 108, 121, 129
Dead reckoning, 102
Defense, U.S. Department of, 215
Defense Industry Productivity Program (DIPP), 88–89, 97
Defense markets. *see* Military markets
Delco Corporation, 104, 173
Delivery schedules, 124–125
Design and production, 7, 26, 59–60, 81–82, 83, 97, 107, 119, 120, 159–160, 169–170, 177, 209–211, 213–217, 232–234
   AN/ARN-152, 237
   CMA-719, 34
   CMA-734, 93–94
   CMA-771, 187–188
   costs, 82–83
   DIPP, 88–89
   failure reduction, 189
   Global Positioning System, 237–238
   Omega system, 17–25
   receivers, 219–220, 222–223, 226–228
   second generation equipment, 71–75, 82
   variations in product, 163
Deveau, Tony, 154, 167, 199, 200, 210
Digital data link, 233–234
DIPP. *see* Defense Industry Productivity Program (DIPP)
Discretionary funds, 225, 226
Distributed processing, 117–118
Distributors, 143–144, 145–146
Diversification, 117, 178, 235
Domestic carriers, 126
Doppler effect, 213
Doppler Navigation System, 5, 19, 44, 48, 58, 63, 66, 80, 90, 94, 100, 104–111, 128, 138, 139
Douglas Aircraft Company, 25, 173, 176
Duplantie, Jean, 150
Dvornik, Yanka, 234
Dynell Electronics, 48, 71–73, 78–80, 90, 94, 102, 103, 105, 110, 111, 113–114, 128, 132, 166

Earthquakes, 67
Eastern Airlines, 101, 124–125, 166
Edison, Thomas, 3, 59
EH101 Heliliner, 236
Electra aircraft, 34, 47
Electrical interference, 23, 63–64, 75
Embraer, 56–57, 122, 129
Emergency service, 189
Engine Health Monitor, 209
Engineering, types, 7
English Electric Company, 4
Entitlements, 224
Equator, 174
Europe, 154, 156
Exports, 8

Factory orders, 182–183
Falcon Jet Corporation, 121
Falkland Islands conflict, 137
Fares, Sohel, 218
Farnborough Air Show, 30–31, 38, 134, 169

Federal Aviation Administration (FAA), 43, 53, 89, 97–98, 173, 209, 215, 233
  certification guidelines, 126, 238
  CMA-2100 SatCom, 238
  CMA-740 approval, 114–115
  equipment approval, 167
  flight tests, 47–48, 57, 102–103
  seminars, 163
  specifications, 93
  use approvals, 108, 114–115
Ferrite core memory, 34–35
Fighter aircraft, 217, 220, 221
Filiatrault, Claude, 139
Finley, Bill, 200
Fixed base operators, 68, 145, 146, 147, 155–156, 199
Fixed-wing aircraft, 73, 120
Flight Advisory Computer, 198–199, 201, 203, 208
Flight inspectors, 163
Flight management systems, 172, 195, 201, 208–209, 211, 232, 235, 236
Flight tests, 43, 45–50, 59, 63, 65–67, 73, 117, 124–125, 134, 168, 216
  accuracy of, 35, 105, 122, 132–133, 159–160, 190–191
  Air France, 169
  China, 207
  CMA-719, 34–35, 57, 63–67
  CMA-740, 132
  joint system, 175
  MD-11 aircraft, 236
  MLS receiver, 226
  P-3 aircraft, 45, 47
  Pan American World Airways, 102–103, 105–106, 121
  requirements for, 167
  Royal Aircraft Establishment, 21–22, 35
  transatlantic, 114
  United Airlines, 167
  Varig Brazilian Airlines, 122–123, 132, 139, 159–160
Ford, Eric, 210
Ford, Henry, 29, 77
Fournier, Pierre, 65, 105, 120, 139, 148, 150–152, 163
Franklin, Benjamin, 143
Fuel management systems, 198–199, 209, 231, 232

Gagnon, Frank, 41
Galipeau, Michel, 11, 12, 21, 108
Gasser, Peter, 11, 20, 22, 32, 59, 60, 74–76, 78–79, 92, 94, 100, 103, 104, 107–109, 114, 115, 117, 124–125, 145, 147–148, 150, 155, 198–199, 203, 207–208, 210
General aviation markets, 73, 98, 110, 119–121, 143, 145, 194–195, 199–200
General Electric Company, 4, 174
Gershanoff, Hal, 95
Gibbs, Christina, 29, 31–32, 57, 100, 202, 208
Gilbert, Rob, 17
Glegg, Kieth, 5–6, 8, 10, 17–19, 32, 34, 43–44, 48, 52, 61, 68–69, 79, 81, 83, 92–93, 117, 148
Global markets, 77, 82, 101, 124, 126, 129, 131, 133, 137, 153–154, 156, 166, 168–169, 194, 199–200, 203, 207–208, 238
Global Navigation Satellite Sensor Units, 237
Global Navigation Systems Incorporated, 40–41, 80, 104, 143, 146, 167
Global Positioning System, 200, 232, 235, 236, 237
GNS-100, 39–41
Great Wall of China, 205–206
Greenland effect, 16
Greenlandair, 131, 194
Ground stations, 4, 46–50, 75, 78, 98, 105, 110, 114, 132, 159, 166, 214, 215, 220, 221, 226, 227
Group managers, 7
Grumman floatplane, 184
Guarantees, 97, 122, 140, 190
Gulfstream aircraft, 121, 202, 209
Gyles, Colin, 22–23, 35
Gyro-based systems, 62, 171, 187
Gyroscopes, 19

Haberl, John, 214–215
Hale, Edward Everett, 87
Hanes, Carl, 68–69, 106–107, 117–121, 126, 143, 145, 147, 167, 200, 216, 217, 223
Hanley, David, 199, 200, 208
Hapag-Lloyd, 194
Harvard University, 6, 12
Hawaii, 15, 59
Hawker Siddeley Aircraft, 108, 129, 131
Hazeltine Corporation, 215

Helicopters, 81, 120, 167, 168, 195, 209, 236, 238
Helms, Lynn, 232
High-frequency communications systems, 233, 234
Hollingsead, Robert, 163–164
Hollingsead International, 163–164
Holmes, Oliver Wendell, 51
Honeywell, 176, 187, 237–238
Howarth, Barry, 139
HS-748 turboprop aircraft, 129

Ice, 16
Ilyushin IL-62 aircraft, 101
India, 204, 205
Indian Air Force, 29
Industry and government, 87–99, 103
Inertial Navigation Systems, 13, 19, 24, 33, 34, 39, 42, 47, 50, 57, 58, 66, 92, 100, 104, 117, 128, 132, 149, 159, 171–176
Inertial-gyro navigation, 62
Institute of Navigation (U.S.), 37, 95
Instrument Landing System, 214, 220, 237
International Civil Aviation Authority, 89–90, 98, 214, 233
International Omega Association, 45, 95–96, 107, 188, 210
Inventories, 183
Ionospheric disturbances, 15, 23, 35–36, 63–64, 136, 159, 190
Italian Air Force, 217

Japan, 15
Japan Airlines, 58, 67, 101, 123, 133, 137, 138, 141, 156, 168, 169, 194
Joffrey, Stephen, 229, 230
Joint ventures, 174–175, 194, 217, 220, 238

Kalman filter, 36, 45, 46
Kanata, Canada, 196–198, 209, 233
KC-135 aircraft, 47
Kelly, Russell, 120, 148, 150, 151
Kelly, Tom, 68–69, 106–107, 117, 119, 120–121, 147
Kemp, George, 4
King, Bill, 103, 105, 107–108, 109, 115, 188
King, Mackenzie, 80
KLM Royal Dutch Airlines, 58, 91, 100–101, 104–105, 113, 116, 166

Kokozinski, Tom, 69, 146
Koller, Hans, 150, 190

L-1011 aircraft, 124, 173
Lackner, Fred, 95
Laker Airways, 66
Lan Chile, 52–53, 123, 194
Land navigation systems, 177–178
Landing systems, 198, 209, 214, 216–230, 235
Lane slippage, 22–23, 35
Lanoue, Jean-Claude, 20, 32, 72, 74, 79, 155, 196, 211
Laser technology, 176
Lavoie, Michel, 20, 32
LCD display units, 235
Leach, Reggie, 157
Learjet, 121
Leasing programs, 179
Lee, Tony, 23–26, 30, 31, 32–34
Lepage, Jean-Pierre, 175, 227
Leveille, Lionel, 108, 110, 128, 145, 200
Liability, 140–141
Liberia, 14, 15, 59, 105
Liberian Airways, 131
License manufacturing, 134, 135
Lincoln, Abraham, 181
Lineas Aereas Paraguayas, 194
Litton Canada, 50
Litton Industries, 62, 71, 80, 90, 103–104, 113, 123–125, 128–129, 132–137, 162, 166, 173
Lockheed Aircraft Company, 27, 44, 49, 108
Lockheed-Marietta, 128, 156, 173, 176
Longfellow, Henry Wadsworth, 131
Long-range high-frequency communication systems, 233
LORAN-A system, 42, 46–50, 52–53, 56–58, 61, 63, 65, 77, 92, 98, 100, 104, 114–116, 158, 168
Lovett, Bill, 120, 139
Lublin, Irving, 45, 46, 95

Mackenzie, Don, 139
Mactaggart, Don, 10–11, 17–19, 21–23, 30–32, 38, 72, 81, 203, 204, 206, 233–234
Magnavox, 14
Magnetic field, Earth, 159
Management and Budget, U.S. Office of, 225

Management by profit, 182–184
Marcel–Dassault Aviation, France, 156
Marconi Avionics (UK), 26, 101, 123, 126, 129, 134, 154, 168
Marconi, Guglielmo, 3–5
Marconi Italiana, 126, 154, 168–169, 217
Marconi Wireless Telegraph Company of America, 4
Marine receivers, 18
Maritime patrol aircraft, 27, 37, 41, 47, 57, 61, 122, 129, 167
Marketing, 24, 32–34, 43, 146, 199–202
  budget, 153, 169
  business aviation, 67–68
  CMA-719, 25–28
  commercial aviation, 98–111
  commitment to customers, 157–164
  customer base, 51–58, 115–116
  department development, 117–129
  equipment loans, 167
  evaluation requests, 116
  exports, 8
  general aviation, 73
  global, 59–69, 82
  initiatives, 165–179
  investment in, 169
  multiproduct, 151
  Omega certification, 115
  personnel, 30, 218
  potential customers, 123
  presentations, 43–44, 52–53, 55–56, 62–64, 204
  representatives, 38, 168
  responsibility for, 6
  sales forecasts, 194
  smaller units, 94
  strategies, 49–50, 60, 103, 104, 121, 126, 134–135, 148–150, 158, 170, 209–210, 218
  tools, 96
  see also specific markets
Matrix organization, 6, 7, 8, 32, 139, 148, 162
McDonnell Douglas Aircraft Corporation, 156, 236–237, 238
MD-11 aircraft, 236–237, 238
MD-80 aircraft, 236
Memory capacity, 75
Michener, James, 3
Microlander, 226
Microwave Landing System, 198, 209, 214, 217, 224–227, 235

Middle East markets, 169
Military communications, 39–40
Military electrical interface circuits, 219
Military markets, 5, 29, 32–33, 37–38, 43, 52–53, 60–61, 63, 95, 97–98, 108, 119, 122, 162–163, 168, 170, 183, 194–195, 197, 199, 214, 217, 235
  NATO number, 26
  product loyalty, 46
  see also specific military organization
Military navigation systems, 232
Miller, Ron, 11, 12, 20, 32, 72
Minto, Lord, 4
Miranda, Antonio, 53–56, 62–63, 121–122, 141
MLS. see Microwave Landing System
Mohr, John, 104, 111, 113
Monarch Airlines, 129, 156
Monitoring and display systems, 232–233
Morris, Peter, 133
Mounayer, Joe, 11–12, 20, 32, 72
Mueller, Arthur, 140–141, 160
Murfin, Tony, 175, 198–199, 201, 208, 215–218, 221, 223, 226, 227, 229–231, 236

National Defense Department, Canada, 216
Naval Air Systems Command, U.S., 37
Navigation errors, 35, 159–160
Navigation management systems, 81, 163, 176
  see also Omega System
Navigation parameters, 20
Navigation standards, 98, 114
Navy, U.S., 57, 78, 198, 215
  communication signals, 39–41
  communication stations, 167
  ground stations, 22
  Naval Electronics Laboratory, 12–13
  Northrop contract, 20, 21, 26, 33, 41, 44–45, 88
  Omega Project Office, 13–14
  receivers, 33, 220
Neudachin, Igor, 203
New Zealand, 14, 15, 31, 106, 133, 138
Niger National Air Squadron, 131
Noise detection instrument, 23
Nordair, 33, 34, 47
North American Engineering Company, 38

North Atlantic Treaty Organization, 26, 43
North Dakota, Omega station, 14
Northrop Corporation, 58, 64, 90, 103, 108
    Air Force, U.S. procurement, 79–80
    AN/ARN System, 37
    Navy, U.S., program, 20, 21, 26, 33, 41, 44–45, 88
    Omega signals, 36–37
    second generation competition, 72
Norway, 14–15, 59

Offshore oil fields, 168
Ohmae, Kenichi, 172
Okura Trading Company, 66–67
Omega Navigation, 163
Omega North Atlantic Minimum Navigation Performance Standards Advisory Circular, 114
Omega System, 10–11, 12–16, 149, 155
    accuracy of, 35, 105, 122, 132–133, 159–160, 190–191
    changes to, 186–188
    CMA-719, 24–28, 32
    components of, 20
    costs, 46
    delivery of, 114
    development of, 17–25
    flight tests. *see* Flight tests
    installation of, 114, 120, 143, 195
    overview, 12–16
    signals, 15–16, 35–36, 39–40
    specifications, 78–79
Omega-Doppler system, 114
Omega-Inertial Navigation System, 117, 172–176
ONTRAC1, 40
Organizational chief, 6

P-3 aircraft, 27–28, 37, 57, 61, 129
Pacific area markets, 133–134, 137–138, 154, 156, 168
Packaging of products, 73–74
Pan American World Airways, 57–58, 91, 99–111, 116, 156, 159, 163, 166, 169, 184
    CMA-740 delivery, 119
    flight tests, 121
    navigation system, 171–173
    negotiations with, 107–109, 181
    Omega installation, 114–115
    service to, 114
Pan-Am/National, 194
Patrol aircraft, 27–28, 37, 41, 42, 43, 57, 61, 129
People Express, 208, 209, 216, 231–232
Performance Data System, 186
Pernase, Eric, 138
Perón, Juan, 52
Piaggio test aircraft, 27
Pierce, John Alvin (Jack), 9, 10, 12–13, 17–18, 22
Polaris submarines, 13
Poldhu, England, 4
Ports of Call Travel Club, 115, 127
Precipitation static, 23
Price, Earl, 50
Prins, Fred, 200
Procurement, 56
    military, 26, 27, 48–49, 61–62, 119
    MLS, 215, 218–230
    Omega system, 77–79
    receivers, 216–217
Product credibility, 64
Product delivery, 119–120, 131, 183
Product development. *see* Design and production; Research and development
Product lines, 4, 6–8, 139, 148–150, 152, 170, 201, 208, 215, 232, 236
    diversification, 235
    financial state of, 183, 184
Product loyalty, 46
Product management, 6–8, 32, 139, 148, 153, 162
Product managers, 32, 72, 83, 134, 148–149, 151
Product support. *see* Support services
Product testing, 21–23, 25, 27, 41–42, 177–178
    see also Flight tests
Production. *see* Design and production
Profits, 118, 181–191
Program managers, 6
Project 2041, 78
Proposal writing, 228, 229

Quality of products, 185–188
Qantas, 138

Radar, 117

Radio Technical Commission for
  Aeronautics, U.S., 80, 91–93, 98, 101,
  106
Radios, 4–5, 12, 150
Raia, Jacques, 103, 105, 107–109, 115,
  121, 173
Rauch, Sol, 81, 196–198, 209, 201, 233
Reagan, Ronald, 208
Receiver-computer units, 20, 25
Receiver-processor units, 74, 93–94
Receivers, 14, 72–73, 75, 77, 107, 167,
  186, 214–215
  commercial, 216–218, 220, 221
  cost, 226
  design, 163
  development, 17–25
  interference with, 23
  military, 216–230
  MLS, 227–228, 232, 237
  signals, 15
  types, 218
  user fees, 42
Recorders, 160
REGA Swissair Ambulance helicopter,
  238
Regulations, 118, 141
Regulatory agencies, 42–43, 176, 200,
  217, 234
  communication signals, 41
  industry and government, 87–98
  see also names of specific regulatory
    agencies
Reliability of products, 97, 189
Remotely piloted vehicles, 178, 193
Repair services, 125, 138–141, 150, 168,
  181, 188, 190, 194
Research and development, 88–89, 97,
  195–197, 201, 210–212
  see also Design and production
Return on investment, 110, 157
Réunion Island, 14, 15
Revenues, 193, 195
  see also Profits
Reynolds, Pat, 103, 114–115, 121, 173
Rieker, Tom, 222
Rio de Janeiro, Brazil, 53–57
Rockwell International, 230, 237
Rockwell-Collins, 201
Rogers, John, 57, 138, 208
Roosevelt, Franklin Delano, 88
Roseberry, Val, 41, 43

Roussel, Gaston, 190
Royal Air Force, United Kingdom, 26–
  27, 57, 108, 133–135, 156, 166
Royal Aircraft Establishment, United
  Kingdom, 21–22, 25, 35
Royal Australian Air Force, 138
Royal Brunei Air, 194
Royal New Zealand Air Force, 138–139

Sakran, Charlie, 37, 45
Satellite communications systems, 214,
  235, 238
Satellite-based navigation systems, 177
Sayegh, Sylvia, 175
Sayegh, Tony, 11, 152, 200
Schlachta, Henry, 96, 117, 118, 126–128,
  131, 143, 167, 199, 208, 210, 235
Scott, Warren, 38
Sea Bee Air, 184
Search and recovery aircraft, 176
Sensors, 176, 195, 211, 236, 237
Separation standards, 89–90
Sercel, 66
Service centers, 194
  see also Repair services
Shaw, George Bernard, 165
Shipborne aircraft, 215
Short-range ship-to-shore radio equip-
  ment, 4
Should-cost analysis, 228
Signal lanes, 35, 36
Signals, 107, 167, 186
  corrections for, 136
  processing, 102–103, 122, 132–133,
    160–161, 173–174
Signal-to-noise ratio, 78–79
Simons, John, 83, 107, 108, 115, 118, 155
Skin mapping, 23
Software, 15, 34–35, 60, 73, 75, 102, 163,
  186
  Kalman filter, 36, 45, 46
Solar explosions, 16
South America, 133, 154, 156, 160
  see also Varig Brazilian Airlines
Soviet Union, 169, 203, 207–208
Space Shuttle Remote Manipulator
  System, 8
Spare parts, 141
Specifications, 91–93, 103, 106, 118, 158,
  164, 188, 219
  ARINC-743A, 238

CMLSA, 228
MLS, 222
receivers, 223–227
Speckled Trout, 67
Sperry Corporation, 62, 187, 201, 217, 219
Spinner, Ervin 107–110, 147, 151, 222, 231
St. John's, Newfoundland, Canada, 4
Static electricity, 23
Status Display System, 199, 201–202, 209, 216
Sterling Airways, 27, 33, 34, 47, 57, 66
Stinson, George, 150
Strategic planning, 60–61, 144, 198, 221–224, 232–233, 235
Submarines, 13, 14
Sullivan, Dan, 222, 228, 230
Suppliers, 124, 161
Support services, 97, 132, 140, 168, 181, 184, 176, 193–194, 235
Surface-mounted devices, 196, 211
Swanson, Eric, 12–13, 15, 35
Swanson Model, 35

Takeuchi, Sam, 67
Teamwork, overview, 9–12
Teleconferencing, 151
Televisions, 4–5
Telex exchange systems, 117–118, 150, 203–204, 233
Test flights. *see* Flight tests
Thibotot, Ray, 150
Tornado aircraft, 217
Total Quality Management, 7, 239
Tracor Incorporated, 71, 78, 80, 90, 94, 103, 105, 110, 113, 124, 132, 166–167, 168
Training programs, 120–121, 143–152, 189
Transport aircraft. *see* C-130 transport aircraft; C-141 transport aircraft
Transport Canada, 93, 234
Transportation, U.S. Department of, 42–43
TransWorld Airlines, 57–58, 91, 100–101, 104–111, 116, 166
Trau, Peter, 89
Travel clubs, 115, 127

Trinidad, 14, 15, 59, 127–128
Turboprop aircraft, 143
Twain, Mark, 71
Twin Otter aircraft, 129, 131

United Airlines, 91, 167, 237–238
United States
    budget process, 219, 224–227
    Omega stations, 14, 15
    see also Military markets
United Technologies Corporation, 99
Universal Navigation Incorporated, 235
Updike, John, 185
U.S.A. National Business Aviation Association, 169
U.S. Congress, 49, 214, 225–227
U.S. House of Representatives, 225
User fees, 42

Varig Brazilian Airlines, 55–56, 58, 63–65, 121–122, 156, 159–161, 168, 169, 173, 190–191, 193–194
    flight tests, 122–123, 132–133, 139
    negotiations with, 139–141
Very low frequency communication signals, 12–13, 15, 20, 39–40, 79, 107, 174, 186
Video conferencing, 104

Walcott, Hank, 95
Walford, George, 120
Warranties, 97, 109–110, 140, 184, 188
Waypoints, 117, 177, 187
Werner, Al, 200
Westland Helicopter, 236
Wilcox Electric, 215
Wilson, Earl, 113
Wireless Telegraph and Signal Company, 4
Wiring harnesses, 162
Woodhouse, David, 21–22
Woodhouse computer, 36
The World Airline Suppliers Guide, 96–97, 109, 183
World Airways, 45
Wright, Frank Lloyd, 213
Wright–Patterson Air Force Base, Dayton, Ohio, 26, 33, 37, 48–49, 218, 219